ACPL ITEM
DISCARDED

S0-BWT-590

333.73
SFEIR-Y
LAND AND SOIL MANAGEMENT

DO NOT REMOVE
CARDS FROM POCKET

ALLEN COUNTY PUBLIC LIBRARY
FORT WAYNE, INDIANA 46802

You may return this book to any agency, branch,
or bookmobile of the Allen County Public Library.

DEMCO

Land and Soil Management

Land and Soil Management

Technology, Economics, and Institutions

Alfredo Sfeir-Younis
and Andrew K. Dragun

Westview Press
BOULDER • SAN FRANCISCO • OXFORD

Oxford & IBH Publishing Co. Pvt. Ltd.
NEW DELHI • BOMBAY • CALCUTTA

Allen County Public Library
900 Webster Street
PO Box 2270
Fort Wayne, IN 46801-2270

All rights reserved. No part of this publication may be reproduced or transmitted in any form or by any means, electronic or mechanical, including photocopy, recording, or any information storage and retrieval system, without permission in writing from Westview Press.

Copyright © 1993 by Westview Press, Inc.

Published in 1993 in the United States of America by Westview Press, Inc., 5500 Central Avenue, Boulder, Colorado 80301-2877, and in the United Kingdom by Westview Press, 36 Lonsdale Road, Summertown, Oxford OX2 7EW

Published in 1993 in India by Oxford & IBH Publishing Co. Pvt. Ltd., 66 Janpath, New Delhi 110 001

Library of Congress Cataloging-in-Publication Data
Sfeir-Younis, Alfredo, 1947–
Land and soil management : technology, economics, and institutions / by Alfredo Sfeir-Younis and Andrew K. Dragun.
p. cm.
Includes bibliographical references.
ISBN 0-8133-8733-7
1. Soil erosion. 2. Soil erosion—Economic aspects. 3. Soil erosion—Developing countries. 4. Soil conservation projects—Evaluation. I. Dragun, Andrew K. II. Title.
S623.S43 1993
333.73'09172'4—dc20 93-4990
CIP

ISBN 81-204-0829-2 (India)

Printed and bound in the United States of America

The paper used in this publication meets the requirements of the American National Standard for Permanence of Paper for Printed Library Materials Z39.48-1984.

10 9 8 7 6 5 4 3 2 1

This book is dedicated to
Alfredo Alejandro, Maria Jose, Maria Francisca,
Natasha Ann, Petrina Jane, Asher Michael and Haakan Alastair.

Contents

Tables and Figures

Tables

Figures

Acknowledgments

Several people have contributed to this book. The background paper prepared by Albert Klingebiel, Consultant, was extremely instrumental in drafting Chapters 4 and 5. Of particular importance is the contribution of Professor Bruce McCarl; several chapters in Part Three draw heavily on his background paper.

A draft version of this book was reviewed and revised by Professors Norman Hudson, Michael Stocking and John Dixon. It is clear that Chapters 2 and 10 were greatly improved by the suggestions made by Professor Michael Stocking. John Dixon made very useful revisions to several sections of this book and prepared two illustrations presented in Part Three.

Several national and international agencies have demonstrated a great deal of interest and have contributed in many ways to this book. The Food and Agriculture Organisation (FAO), the World Bank/FAO Cooperative Program, the World Food Programme, the World Meteorological Organisation and UNESCO deserve special mention. These agencies supplied extremely useful materials and many of their professionals have given ideas that proved to be extremely valuable.

Several research assistants helped review the literature and put together large data sets. Our sincere appreciation to Christopher Philippi, William McGrath, James Flemming and Jean Curling.

Drafting this type of book required a comprehensive review of the literature. This review could not have been done without the assistance of the Bank's Library staff. Final preparation of the text was completed by staff at LaTrobe and we thank Bev Tannock for the preparation of many of the tables and Bjorn Jakobsson, who toiled through the final draft.

It is clear that without the support from the technical advisers and colleagues in the Economics and Policy Division, this book would have been much more difficult to write.

Alfredo Sfeir-Younis
World Bank

Andrew K. Dragun
LaTrobe University

The World Bank does not accept responsibility for the views expressed herein, which are those of the authors and should not be attributed to the World Bank or its affiliated organisations. The designations employed, the presentation of material and the maps used in this document are solely for the convenience of the reader and do not imply the expression of any opinion whatsoever concerning the legal status of any country, territory, city, area or its authorities, or concerning the delimitations of its boundaries or national affiliations.

PART ONE

1

Introduction

The chief purposes of this book are to create greater awareness of the magnitude of soil erosion and land management in developing countries, provide analytical frameworks for the preparation and appraisal of soil erosion control programs and discuss the broad range of important issues involved in the establishment of such programs.

The book has five parts. Part One is this introduction. Part Two discusses the physical and economic dimensions of soil erosion and estimates the overall magnitude of the soil erosion problem in developing countries. Part Three focuses on how to evaluate soil conservation programs by integrating scientific knowledge with economic methods and procedures -- and concludes with a series of illustrations of how different evaluation methods have been used. Part Four outlines the most important organisational, institutional, technical and macroeconomic issues involved in appraisal, presents the major elements of project or program design and suggests subjects for future research. The final part is the extensive bibliography.

Nature of the Erosion Process

Soil Erosion

Soil is one of the chief natural resources used in agriculture. Maintaining and if possible, increasing the productivity of existing soils is necessary to accomplish production and welfare goals in both the short and the long term. Despite its undisputed importance as a natural resource, soil is now suffering degradation of various kinds at such a high rate that future agricultural development may be severely limited. Soil erosion is the commonest form of land degradation today, but soil is also being degraded by such things as high salt and alkali levels, deposits of radioactive and inorganic wastes and saturation by chemicals.

Soil erosion is a three-stage process, the removal of soil particles (detachment), the transportation of these particles and their deposition in other areas. The chief agents of erosion are water and wind. An understanding of the entire process is essential not only for scientific purposes but also for understanding the costs and benefits of soil conservation programs.

The most important types of water-caused erosion are sheet erosion, rill erosion, gully erosion and stream bank erosion. Sheet erosion is the removal of thin layers of soil by raindrop splash and the subsequent flow of water over relatively flat surfaces. Rill erosion occurs along soil channels (that is, rills) that are small enough to be obliterated by normal tillage operations. Gully erosion, on the other hand, occurs in soil channels that are too large to be eliminated by ordinary tillage. Stream bank erosion occurs when soil is detached by water in permanent streams.

Erosion affects the potential productive capacity of soil by altering the medium for plant growth. Generally, the topsoil that is lost by erosion provides the most hospitable medium for root growth, water retention and nutrient storage. The erosion process is often difficult to detect by eyesight, it is selective and it is nonuniform. It is difficult to detect because erosion tends to occur slowly and reductions in productivity often manifest themselves only when serious damage has already occurred. It is selective in that the removed soil often contains more nutrients than the remaining soil, which has greater proportions of sand and clay. It is nonuniform in that it can result in a surface seal, or crusting, making farming difficult. This non-uniformity can lead to spotty crop maturities.

Erosion is a natural process. There is no doubt however, that human actions can have a strong effect on the process. These actions include deforestation, certain forms of intensive agriculture, unwise agricultural practices, shifting cultivation, overgrazing by livestock, forest fires, population resettlement and inadequate land tenure systems. Erosion is also exacerbated by human activities that take place outside the agricultural systems (for example, road construction).

To maintain soil productivity at acceptable levels, conservation practices must be carried out. These practices fall into two broad categories, biological and mechanical practices. Examples of biological practices are changes in crop rotation, afforestation, strip cropping, establishment of wind breaks and sand dune stabilisation. Examples of mechanical practices are construction of bench terraces, stormwater drains, artificial watercourses, contour bunds, ridges and furrows.

A Typology of Soil Erosion Effects

The physical nature of the erosion process provides the basis for a typology of erosion effects. Most of the effects of erosion are either "upstream effects," "downstream effects," or "worldwide effects." Little is said in this book about the last category, although recent research shows that erosion has important intercountry and intercontinental effects which policy makers should take into account.

Upstream effects are chiefly the effects of erosion on agriculture within a project area (however this is defined) or in the upper portion of a watershed, while downstream effects are effects on activities elsewhere in the economy. Examples of upstream effects are on-farm losses in productivity and inter-farm damages to irrigation terraces, roads, bridges and other capital assets. Examples of downstream effects are sedimentation of rivers, siltation of reservoirs, floods, contamination of drinking water supplies, constraints on water navigation and hydroelectric power production and damage to environmental services and activities (for example, fisheries and wildlife).

Magnitude of the Erosion Problem

Erosion is very widespread and the upstream and downstream effects of erosion are reaching alarming levels in the developing world. While the data are far from complete, statistics at the national and local levels are comprehensive enough to support the assertion that soil erosion should be an issue of public concern at all levels of decision making in both developed and developing countries. If erosion continues, we can expect a constant diminution in available land assets.

This book groups and analyses data on upstream and downstream effects. Because there are very large numbers of effects, both upstream and downstream, a subset of effects was analysed. The upstream effects include farm-level soil erosion, water runoff and losses in plant nutrients. The downstream effects include sediment loads in major river basins, the effects of sedimentation on water reservoirs and the effects of floods on human lives, livestock and infrastructure.

Erosion at the Country Level

Estimates of the extent of erosion in a number of developing countries are given in Part Two. It is estimated, for example, that land degradation affects nearly 150 million of India's 328 million hectares of land. Of this, 90 million hectares are affected by water (rainfall) erosion. An additional

7 million hectares are seriously affected by salinity and 20 million hectares are affected by floods, some of which can be traced to erosion effects. It has also been estimated that farmers in India are losing 6 million tonnes a year of nitrogen, phosphorus and potash due to the high rate of erosion. Annual replacement of these nutrients would require an expenditure of up to approximately US$6 billion at 1984 prices. Soil erosion is an even more critical problem in certain other developing countries. In Guatemala, for example, 40 percent of the production capacity of the land has been totally lost due to erosion. Farmers in several regions of Guatemala have abandoned their farms because it has become uneconomic to carry out agricultural activities. In Nepal, where many areas have been cleared for subsistence cultivation, losses of topsoil average 35 to 75 tonnes per hectare per year. Erosion in some gullied areas is reported to range from 200 to 500 tonnes per hectare per year. Given the "threshold level" of 5 to 10 tonnes per hectare per year (as established in the United States), erosion rates in Nepal are between seven and one hundred times higher. The World Food Programme reports that more than 2,000 tonnes of soil per square kilometre are lost every year in Ethiopia. This soil, otherwise, would be sufficient to produce food for 12,000 families.

Erosion and Land Productivity

Many factors determine land productivity and research studies that show a clear and convincing relationship between erosion and productivity are few. The FAO Agricultural Department is now conducting an extensive study of this particular question. This book presents evidence on how erosion rates are affected by crop management and other environmental conditions, on how erosion depletes nutrients in the soil and in a few instances, on the effects of erosion on crop yields. Reference is also made to the apparent effectiveness of soil conservation practices in sustaining productivity.

This book shows that erosion rates in developing countries are often very high and that the amount of plant nutrients lost because of erosion is extremely large. Consequently, there are bound to be important effects on land productivity. A few studies have been carried out to determine the relationship between changes in soil depth and crop yields. One study in the United States showed that a loss of 10 inches of topsoil could reduce corn yields by nearly 40 percent.

This book also presents data on the effectiveness of specific erosion control practices. It is clear that, other things being equal, these practices can greatly reduce erosion rates. The acceptance and success of these practices depend on economic, social and environmental conditions.

Downstream Effects

The book focuses first on sediment loads in the major river basins of the world. Except for geologic erosion, these sediments are "produced" upstream by many economic activities, of which agriculture is only one. Sediment loads were converted into an approximate average erosion rate for each basin where data were available. The data systematically show that erosion rates in these river basins are very high compared with "normal" rates and that the rates are independent of the size of the drainage area. The data show that the highest erosion rates are registered in Asia.

Much of the sediment resulting from soil erosion is deposited above major dams or in river beds, often causing severe economic and environmental damage. This book includes data on estimated and actual rates of sedimentation for a large sample of major dams in developing countries. These data show that actual sedimentation rates are several times higher than the rates that had been predicted prior to dam construction. In some cases the actual rate has been more than 20 times the estimated rates. In other words, water reservoirs are literally becoming soil traps. A study of nearly 70 dams in India estimated that the total capacity already lost due to sedimentation was more than 21% Thus, the economic life of very expensive infrastructural assets is being impaired at a rather fast rate. Several other examples of high sedimentation rates are given on a country-by-country basis.

The book contains a scenario analysis of what would happen to reservoir capacity by the year 2000, using different assumptions about sedimentation rates. The first scenario assumes that the "live" storage areas of major reservoirs have not been affected since 1940 and will not be affected between now and the year 2000. The second scenario assumes an across-the-board sedimentation rate of 2% of live storage per year. The third scenario assumes a 1% rate of sedimentation per year for the 1940-50 period, a 2% rate for 1950-60, a 3% rate for 1960-70 and a 4% rate for 1970-80. In the case of the second scenario, it is estimated that the world will lose nearly one-third of total reservoir capacity by the year 2000. The third scenario would mean that two-thirds of total reservoir capacity would be lost by the year 2000.

The data show that while the number of dams increased in earlier decades, it decreased slightly between 1970 and 1980. This may indicate that the number of potential dam sites has declined. On the other hand, the data on total capacity in the different decades show that total capacity has been growing since 1940. This may mean that countries are building dams of greater capacity.

The facts have great implications for the design and implementation of sediment management schemes. In particular, they suggest that where only a small part of the drainage area of a river basin is located in a given country, efforts to control the effects of sedimentation will require action by other countries bordering the rivers.

This book also attempts to estimate an average erosion rate for developing countries. On the basis of data on the sediment loads of rivers and on various analytical assumptions, the book concludes that, on average, developing countries are losing 53 tonnes of topsoil per hectare every year, or more than five times the threshold level of 10 tonnes per hectare per year. Some of this topsoil is deposited elsewhere in the watershed, causing siltation problems, but much is irrevocably lost to the ocean. Soil erosion rates vary from one continent to another. In Asia, soil losses average more than 138 tonnes per hectare per year, compared to 6 tonnes per hectare per year in Africa and 12 tonnes per hectare per year in Latin America.

The emphasis in this book is on the negative effects of erosion and sedimentation. It is important to note however, that these natural processes may also have beneficial effects. Certain ancient civilisations built dams for the sole purpose of trapping sediments. Once the reservoir area was filled with sediments, often with a very high nutrient content, the area was used for cultivation. Some of the most fertile valleys in the world owe their fertility to sediment deposition from upstream.

Economic Nature of Soils

Part Two of this book includes a chapter that discusses the economic nature of soils. This chapter develops a framework for resource management policies, outlining the meaning of conservation and the nature of policy decisions on conservation. The book suggests that the "Safe Minimum Standard of Conservation" (SMSC) should be a guiding policy principle. Given the class of natural resources under consideration -- that is, renewable resources with "critical zones" -- the SMSC is a standard that avoids irreversible damage to the resources.

In assessing the merits of decisions on soil erosion control, a distinction is made between private or farmers' decision making and national or social decision making. While farmers will make decisions which tend to maximise their profits, society's decisions should involve maximisation of the welfare of present and future generations. Farmers make their decisions on an environment where economic factors are usually changing. Some of these are market factors (for example, prices, taxes, subsidies), while others are nonmarket factors (for example, tenure, property rights). Changes in this institutional environment may emphasise either conservation or depletion of soil resources.

Investment decisions are the central topic of this book. It is often contended that the rationale for using Benefit-Cost Analysis (BCA) to make decisions on soil conservation cannot be the same rationale for using BCA to judge other investment projects. What makes soil erosion control projects so unique that they require new or modified BCA procedures? The following reasons seem to be important.

1. Soil conservation decisions have an impact on equity (farm equity compared to equity in other enterprises) and an impact across generations.
2. These projects also have external effects (outside project areas).
3. Valuation of such projects is difficult because of the importance of nonmarket factors.
4. Such projects have very long-term effects.
5. The risk and uncertainty of such projects is large.
6. Soil conservation programs are multiproductive in nature (thus, forestry programs are often designed to control erosion as well as to supply fuelwood energy).
7. Economic evaluation is affected by the presence of ecological irreversibilities.

Economic Evaluation of Soil Conservation Programs

Economic analysis of soil conservation programs often suffers from three shortcomings. These relate to:

1. The identification of benefits and costs.
2. The valuation in monetary terms of goods and services that are not directly traded in markets.
3. Discounting procedures.

Part Two of this book should provide appraisal teams with enough material to identify the benefits and costs of soil conservation projects. Part Three focuses on the problem of valuation, while the problems associated with discounting procedures are discussed in one of the chapters of Part Four.

The quality of the valuation process depends upon the ability of the economist to understand the technical relationships necessary to conceptualise the physical effects of erosion and on proper definitions of the environmental effects of erosion. An adequate system for the valuation of benefits and costs must take account of the principles of welfare economics, particularly the notions of consumer surplus and rent. Part Three of the book presents the most important principles of project evaluation and also the corresponding procedures for estimating the necessary parameters to be used by the economist to assess changes in welfare "with" and "without" a soil conservation project (changes in both upstream and downstream effects).

Notwithstanding the limitations imposed by a lack of data, an attempt is made to integrate scientific frameworks with economic methods and procedures, as well as to extend traditional Benefit-Cost Analysis. The book approaches the problem of integrating scientific knowledge with economic methods in the following way. First, it describes such natural phenomena as erosion, sedimentation and floods. Second, it illustrates the technical relationships of greatest interest to natural scientists (for example, the relationship between floods and soil erosion and the relationship between soil erosion and sedimentation). Since the analytical frameworks respond to environment-specific variables and circumstances, a few predictive models are shown for illustrative purposes.

The book emphasises the importance of linking technical frameworks with changes in the production of tradeable commodities. In particular, it is proposed that environmental goods (for example, soil) should be desegregated into their characteristics (that is, size, shape, volume, chemical nature) and that changes in these characteristics can be related to changes in productivity. Instead of concentrating on losses of topsoil, the book suggests that changes in the characteristics of the soil left on the ground (for example, availability of nutrients, soil depth) should be examined to determine how they affect crop yields. Similarly, it is suggested that sediment should be desegregated into such characteristics as volume, particle size, particle shape and chemical concentration and that changes in those characteristics should be related to the profits or costs of hydro-electric power, downstream irrigation water, drinking water, riverine transportation and so on.

If the data and models used by natural scientists are unrelated to productivity, it will be very difficult to improve economic evaluation methods. Several examples are given in the text to illustrate this point.

To improve the process of monetary valuation, the book proposes assessing changes in tradeable commodities (for example, energy, water, transport) brought about by soil conservation projects. The valuation process is done by examining changes in net profit (income), total cost, marginal cost, input supply and others. As stated earlier, the key to using the proposed methods is to assess the extent to which they agree with the principles of welfare economics. The theoretical formulation that enables analysts to appraise projects through changes in profit levels is called "duality theory."

Technical Frameworks

The main purpose of one section of the book is to illustrate some of the most important technical relationships and models used in quantifying the physical effects of erosion upstream and downstream. Examples of approaches to quantifying and predicting soil losses, nutrient deficiencies, floods and sedimentation are presented in the text. The idea is not to cover the literature exhaustively, but to illustrate the types of relationships that are the focal point of many technical models. The output of these models will greatly determine the extent to which economic appraisal can be carried out. One of the predictive models which is examined in some detail is the Universal Soil Loss Equation.

Even if a framework described here cannot be applied in the short term because of lack of adequate data, the description and accompanying analysis will enable appraisal teams to design appropriate monitoring and data collection procedures.

Economic Frameworks

A large number of economic frameworks are presented to permit appraisal of both the upstream and downstream effects of soil erosion in fiscal terms. The quantification of economic effects could be carried out through estimates of changes in net revenue or profits, changes in total or marginal costs, or changes in input requirements and output supply. The main evaluation methods are methods based on observed economic behaviour and methods based on synthesized (or simulated) economic behaviour.

This book presents several procedures for judging the economic merits of soil conservation programs, procedures that in some cases are extensions of traditional Benefit Cost Analysis. Some of the procedures focus on the benefit side, while others focus on the cost side. Examples of methods dealing with the benefit side include direct application of the consumer/producer surplus approach, the land value approach and the income foregone approach. Examples of methods dealing with the cost side include the preventive cost approach, the replacement cost approach, the least cost approach and the shadow project approach.

The correspondences between each approach and the basic principles of welfare economics are explained. In the least cost approach, for example, the analyst assumes that a goal or target level (or soil erosion, of water runoff) is given and that several different soil conservation programs could be used to reach this goal. The analyst then tries to determine which program has the lowest cost and no attempt is made to quantify benefits. Other approaches estimate the benefits of a program and then compare those estimated benefits to the costs of the program.

Several procedures are presented in detail. These procedures take into consideration the following -- purpose, problem under examination, technical data requirements, other data needs, application methods, type of estimate generated, most significant assumptions, ways of generating total benefit/cost estimates, references to case studies and comments on their use.

Illustrations are presented to show how some of the procedures have been applied. Since the parent materials were not originally written in case study form, the presentations in some instances are rather limited. After reading this material however, the analyst should be able not only to understand the merits of each approach but also to modify the approaches to fit specific circumstances.

Major Issues and Future Research

Project Appraisal Issues

The most salient technical issues include determining the effects of erosion, relating erosion to crop yields, nonuniformity of effects, quantifying sediment delivery on land, assessing the impacts of soil erosion on water, the episodic nature of erosion, the effectiveness of soil conservation practices, analyzing factors that mask erosion effects, defining attribution factors and availability of data. The thrust of this section is on reconciling the quantification of erosion effects in physical terms with the need to identify indicators of welfare effects.

Two examples may serve to illustrate these concerns. Much of the work on soil erosion is intended to quantify the amount of soil that leaves farm fields. Although this measure gives some indication of the magnitude of erosion and its potential impacts downstream, the amount and characteristics of the soil that is left behind are more important in appraising a project's impact on farm productivity. Another example concerns the episodic nature of erosion. While most data come in the form of averages (tonnes/ha/year), it is well known that the factors affecting erosion rates vary. This variability is very important in selecting the type of method needed to control erosion. If economic analysis of soil conservation programs is to improve, data collection should take into account the episodic nature of erosion and the corresponding effects on productivity.

The most important economic issues in appraisal are the integration of technical and economic measures, the scope of the appraisal effort, the *development of practical measures* and ways to deal with "unprofitable" projects.

Organizational and Institutional Issues

The organizational and institutional aspects of soil conservation programs are of great importance. Technical solutions are often available, but potential benefits never materialize because of organization and institutional deficiencies.

There are a number of important organizational issues. The first is the proper role of government, not only as an important investor in conservation projects but also in regard to planning and budgeting, information gathering and dissemination and project direction at the national and local levels. The second is the issue of research into agricultural matters necessary to improve the assessment of conservation projects. The third is the need to emphasize the importance of conservation programs, as opposed to production-oriented programs. The fourth is the need for better monitoring and evaluation management instruments. The fifth is the importance of farmers in the planning and execution of soil conservation projects, it is the farmer who will determine the success of a conservation strategy, either by accepting it and implementing it enthusiastically, or by ignoring and thus undermining it. Related to this is the issue of how to determine the acceptability of any particular strategy. A sixth important organizational issue is achieving better coordination among international donor agencies both within individual countries and across political boundaries. A last organizational issue is the development of ancillary activities, such as the marketing and distribution of agricultural inputs necessary to improve soil conservation.

The most important institutional issues involve tenancy, property rights, credit arrangements and laws. Incentives affecting the conservation of soil are greatly affected by tenancy. If farmers are not assured of rights on the land they cultivate, there will be no incentive to invest in such long-term programs as soil conservation. Different forms of property rights also have effects on soil conservation. Efforts to expand livestock production, for example, have often reduced the carrying capacity of lands held in common, such as grazing lands. Because of the long-term nature of soil conservation investments, creation of liquidity through credit has become an important instrument of soil conservation programs. However, problems often arise when these credit programs are not targeted properly. Finally, the formalisation of incentives and land protection laws is important and in this context enactment of legislation is a crucial element. However, enforcement of incentive programs and land protection laws is sometimes prohibitively expensive.

Macroeconomic Issues

The most important macroeconomic issues are the establishment of priorities and standards (that is, how much of its soil conservation problem does a country want to solve), the codification of resource management practices, the need to integrate conservation policies with income policies ("poor people in poor lands"), the "energy dilemma" that arises because so many rural people satisfy their energy needs with fuelwood, shifting population movements, the effects of soil conservation policies on income and wealth, the establishment of investment priorities and large-scale population resettlement.

Incentive Programs and Intergenerational Equity

Compensating farmers or other groups as a reward for adopting potentially unprofitable conservation practices is advocated by many decision-makers. The operation of such incentives as price supports, subsidized interest rates, general subsidies and direct compensation occupies a large part of the literature on conservation in both developed and developing countries. How policy makers arrive at an optimal compensation package, in cash or in kind, is important, but analysis of the foreseeable micro and macroeconomic impacts of incentive programs is seldom carried out.

Several compensation schemes are presented and discussed with respect to their effects on the adoption of soil conservation practices, on administrative operations and maintenance, on income and prices and on labor. Lack of proper assessment often means that these incentive schemes inadvertently defeat the purposes they are designed to accomplish. The major issues addressed are what to compensate for, how best to compensate, when to compensate and who to compensate.

This book presents several incentive schemes carried out by the World Food Programme (WFP), which has been assisting soil conservation efforts since 1964. As of May 1984, the program had assisted 43 soil conservation projects and 78 projects with soil conservation components, its assistance amounting to $242.3 million.

Several lessons have been learned from these efforts:

1. Effectiveness in controlling erosion depends on comprehensive land use planning rather than on piecemeal approaches.
2. Limits on the volume and duration of aid should be established.
3. Programs and techniques should be adapted to local conditions.
4. Food aid programs should be targeted to socially and technically desirable projects.
5. Incentive schemes must be accompanied by other interventions (particularly income transfer).
6. Present and future conservation efforts should be sustainable after aid stops.
7. Governments should expand and improve the agencies implementing such schemes.

One of the basic attributes of soil conservation projects is that their benefits do not become evident quickly. Although the costs of these projects are borne by the present generation, their benefits (or the future costs of not conserving soil) will accrue to future generations. The costs of downstream erosion, on the other hand, are often borne by future generations. This issue of present and future benefits and costs raises important questions about the appropriate appraisal procedures for selecting soil conservation projects. An "economic" criterion will differ greatly from a "conservationist" criterion. Further, the economic criterion is less useful in establishing investment priorities that will minimize undesirable equity patterns across generations. One solution lies in a "supra rule" which would condition the way in which intertemporal social choices are considered. While an economic criterion based on standard discounting methods provides useful information and working rules that enable policy-makers to achieve an efficient and equitable distribution of welfare for their own generation, the same criterion often fails in establishing acceptable equity patterns across generations. Ways to reconcile the two criteria are presented.

Strategy and Future Research

The most important elements of a soil conservation strategy will depend upon each individual country's development opportunities and options. Solutions to the issues outlined earlier depend on environmentally specific aspects of country development. Therefore, instead of presenting a specific strategy, the book presents a framework for developing strategies. Any strategy will have three levels of decision-making -- the policy level, the institutional level and the operating level. The policy level focuses on objectives and goals, the institutional level deals with incentives, while the operating level determines the allocation and use of natural resources. Soil conservation programs should be taken into account at all three decision-making levels.

This book recommends that the policy level focus on the establishment of a resource management policy, integration of resource management with overall development objectives, creation of greater awareness of soil conservation's importance, creation of appropriate technology policies, allocation of national funds, formulation of policies on public and private sector participation, consideration of the rights of future generations and formulation of realistic targets and goals. At the international level, this book recommends coordinated action by International Funding Agencies (IFAs), financial resource mobilisation and expansion of sector work and policy dialogue on environmental issues and constraints. At the institutional level, this book recommends the establishment of market and nonmarket incentives and the assessment of the potential impacts of these incentives on the allocation and use of land resources. Finally, at the operating level, this book recommends improvements in the design and formulation of soil conservation programs, expansion of multi-disciplinary assessment teams, unification of existing organisational structures, the creation of monitoring and evaluation units, expansion of conservation-oriented extension services, development of adequate training and research programs, development of ancillary activities and improvements in the marketing and distribution of essential inputs for soil conservation activities.

The book lists a series of topics for further research -- the relationship between soil erosion and productivity, the elements that determine program or project success, assessment of the downstream effects of erosion, regional integration of soil conservation programs, organisational arrangements, the impacts of nonmarket institutions, the creation of data banks, assessment of compensation schemes and the contribution of soils to economic development.

How Did We Get Where We Are Now?

The nature and magnitude of soil erosion in developing countries is such that action is urgently needed. Although erosion is a natural phenomenon, imbalances have been created by human actions. Consequently, erosion and sedimentation will impair the achievement of long-term development objectives. Macroeconomic policies designed to deal with prices, export incentives, foreign exchange earnings and the like, that would appear on their face to have no connection with the degradation of land, have resulted in an array of incentives to use land unwisely.

While an understanding of the economic aspects of soil erosion is essential, controlling or avoiding land degradation is not only an economic problem. Social, demographic, physical and perhaps most important of all, political factors will determine the future success of soil conservation efforts. Investment decisions -- the main subject of this book -- are only part of the dilemma. Decision makers will have to change their present perspective and integrate soil conservation decisions with policies and institutions that deal with food production, research, extension and education.

Assessment is basic. National authorities should ask questions like these: Do we know enough about the present status of our natural resources? Do we know enough about claims to use the country's natural resources and those who make the claims? To address those questions, a decision has to be made to change existing *units of account*. Instead of purely sectoral approaches (for example, food production, livestock, forestry, irrigation, electricity, water supply), comprehensive solutions to these problems will necessitate a "total catchment management approach." A river basin is exploited by many sectors and sectoral subsectoral actions have resulted in tremendous fragmentation. Changes in existing units of account will necessitate important institutional reforms.

Where Do We Go from Here?

Given the seriousness of the soil erosion problem, the worst action would be to do nothing -- to take no risks. Decision makers should realize that they may fail several times before seeing the light of success. A review of the voluminous literature on soil conservation indicates that a range of determinants of success can be clearly identified:

1. Soil conservation practices are not necessarily expensive.
2. Farmers must participate in the decision making process.
3. Management objectives and policies must be developed and expressed clearly.
4. Personal income and food policies must be integrated with conservation practices.
5. Effectiveness at the policy or program implementation level depends upon knowledge and information.
6. Investments and policies should be valued not only on their economic merits but also on other grounds, including ethical or moral grounds.

7. Replicability and area coverage are two basic determinants of program success.
8. Investments and policies will have to be oriented toward internalising the negative effects of human actions, whether they are the actions of farmers or of other actions in the economy.

How Do We Get Where We Want to Go?

An integrated approach across sectors, across decision-making levels and across private and public sector boundaries is essential. Rational exploitation of the natural resources available to agriculture and other sectors will require a clear set of management objectives. Without such objectives, investment and policy changes will be carried out in a vacuum. Several prerequisites to action can be singled out:

1. The need to investigate potential changes in soil and water quantity as well as quality.
2. The need to recognise demographic problems.
3. Careful assessment of conservation/development tradeoffs.
4. Definition of an action program to lead the way into the future.

A basic set of management objectives must be followed to avoid irreversible damages that will result in constraints to future development. It will be necessary to establish cost-effective measures that will reverse soil degradation problems in the medium term, to agree on reducing erosion to tolerable levels and to adopt policies that will encourage the retention of prime agricultural land. Further, investment policies should be designed to minimize the adverse effect of organic wastes in water, to reduce water pollution from excessive nutrients and salinity, to minimise levels of toxic pollutants and to reduce sedimentation.

It will be important to identify areas where populations place stress on the natural resource base. With regard to the energy dilemma, major efforts should be made to deal not only with forestry programs but also with energy conservation technology, energy substitutes and rural energy programs in general. This will mean motivating people to participate in cost-sharing programs, paying for larger research and education projects and building a national consensus.

Coping with the challenges ahead will not be an easy task. Countries and financial institutions will need to establish comprehensive conservation agendas, set reasonable targets and focus on achievements.

Let us remember that poor land makes people poor.

PART TWO

2

Nature of the Erosion Process

Definitions

The Soil

Soil is a dynamic surface material in which complex biological, chemical and physical activities take place. Substances in solid, liquid and gaseous forms (organic and inorganic) are found in the soil. A range of soil types are found around the world. Most of them contain a common set of elements and the major source of differentiation is the relative proportions of those elements. Soils are composed largely of particles that differ in size and shape, the most important being sand, silt, clay and organic matter (living, dead, or decomposed). Soils have different layers, the topsoil (also called plough layer) and sub-soils. The topsoil is often characterized as having more organic matter than the underlying subsoils. The situation is reversed with regard to the proportion of clay content.

Roots require a proper environment for growth as they expand into pores or cracks in the soil. This proper environment is characterized by appropriate temperature, degree of moisture and aeration capacity. Soil texture and structure determine the working properties of a soil, especially water relationships and the availability of nutrients to growing plants.

Soils contain organic matter and nutrients. One gram of soil contains millions of bacteria, their quantity, type and role depending on their soil environment. Nutrients come in varying quantities and in degree of solubility and availability, to plants. Major (or macro-) nutrients include nitrogen, phosphorus and potassium, while trace elements (micro-nutrients) include iron, zinc and boron.

Soil fertility depends on several characteristics, including infiltration of water, available water storage capacity (especially when plant water requirements exceed storage), mineralization (breakdown) or organic matter and temperature.

Changes in soil structure can reduce soil fertility by affecting the availability of organic matter, water and other parameters related to plant growth.

Soil Erosion

While the focus of this book is on soil erosion, it should be remembered that soil is also being continuously created by natural processes. As bare rock is broken down by the action of water, temperature and wind, soil is created and the top layer is mixed with organic matter. The process is very slow, however, a study in Indiana found rates of soil formation from different sources to vary one inch every 25 years to one inch every 500 years. When soil erosion begins however, it usually proceeds at a much faster rate than soil creation, thereby leading to a long-term net loss of soil.

Soil erosion is a physical phenomenon of the soil surface which has economic affects, both on upstream soil quality (and thus on yields) and on amounts of waterway sediment (and thus on water and habitat quality). Erosion is a process that includes three steps, the *detachment* of particles of soil by wind and water from the surface, the *transportation* of these particles and the *deposition* of these particles in another place.

In water erosion, detachment occurs when raindrops strike the soil, or when running water picks up loosened soil and transport it, or when water acts as a lubricant and causes soil to slip down slopes.

The amount of detachment by rainfall is closely related to the intensity or rate of rainfall. The amount of soil that can be washed away by surface runoff depends on the volume of runoff and the velocity of flow. Some of the rain is absorbed by plants and soil, the surplus becoming surface runoff. The amount of runoff is increased where there is a large amount of rain and low absorption, because of poor plant cover or shallow soil.

Soils damaged by erosion thus become more vulnerable to further erosion. The velocity of surface runoff affects its power to cause erosion, so steep slopes or long slopes are more at risk. One of the chief control measures is to decrease steep slopes by bench terracing, or to break up long slopes by horizontal banks that intercept the runoff.

Water erosion has been classified in many different ways. Here the following four classifications are used (following Trolh, Hobbs and Donahue, 1980):[1]

1. **Sheet** [illegible] removal of more or less uniform layers of soil. It is caused [illegible] sh and the flow of sheets of water over

entire surfaces. Sheet erosion may appear to be a slow process because only a few millimetres of soil are removed annually, but relative to total soil removed (tonnes/ha) and geographical extent, it is the most serious form of erosion. One of the worst aspects of sheet erosion is its insidiousness.

2. **Rill erosion** -- erosion occurring along small channels or rills small enough to be obliterated by normal tillage. Rill erosion is common in newly-tilled land, on steep slopes and in such places as building sites and roads embankments. Both detachment and transport of soil occur along the rills.

3. **Gully erosion** -- erosion occurring in channels too large to be eliminated by tillage. Gullies extend by headward recession and waterfall action, as well as by the undermining of their sides by flowing water. Gullies act as major lines of transport out of a catchment for sheet-eroded, rill-eroded and gully-eroded soil -- a high velocity flash flood may do this.

4. **Stream bank erosion** -- erosion of soil from banks by water in permanently flowing streams. Stream bank erosion is not triggered directly by rainfall, but is increased by higher rainfall.

Many factors affect the rate at which erosion takes place. The most important are -- climatic, topographic, pedologic, geologic, morphologic and hydrologic factors. Major climatic factors include temperature, rainfall, sunlight and winds, rainfall being the most important. Temperature determines the extent to which precipitation will become rain, causes snow to melt at a certain pace and therefore determines the rate of runoff, causes soil to freeze at different depths and thus affects rate of infiltration, determines soil moisture and regulates the rate of evaporation. These climatic factors are often affected by the amount of solar radiation. Important topographic factors are slope (for example, degree, length) and soil shapes. Morphologic and geologic characteristics of soils and their permeability, texture and structure all affect the rate of erosion, depending on vegetation (for example, extent, type) and the type of erosion control practices carried out in any specific area.

Humans are an important agent of change, affecting the rate at which soils are being eroded around the world. The most important factors are, deforestation, driven most often by demand for wood-fuel energy as well as traditional forestry products,[2] certain forms of intensive agriculture, inappropriate agricultural practices, such as ploughi[illegible] up-down rather than acoss the slope of a hill,[3] shifting cultivation, [illegible] and is left fallow only for short periods of time, overgrazing,[4] [illegible]

Other factors that deserve mention are poverty (crating pressures to use land continuously to satisfy subsistence needs) and high population density compared with the carrying capacity of existing soils. Important institutional factors are insecurity of land tenure, which makes farmers unwilling to invest in land seen not to be theirs and small holdings that make it impossible to carry out conventional soil conservation practices.

Economic factors are also important. These will be treated in detail in Chapters, 6, 7, 9 and 10, but changes in prices, interest rates, taxes, subsidies and marketing costs are all significant determinants of farmers' attit
udes toward the conservation of soil. These factors are often called "incentives," whether they operate through markets or outside market forces (that is, as state regulations). The traditional way in which the markets provide signals to producers and consumers in their allocation of resources often does not benefit the economy relative to environmental decisions in general and soil conservation decisions in particular, because environmental attributes are not traded. These issues are discussed in detail later.

The Process of Erosion

Economic analysis of soil erosion programs often begins with the establishment of some notion of productivity, commonly crop yields. Erosion alters the characteristics of soils as a medium for plant growth by changing productive capacity. Erosion reduces productivity mainly through loss of water retention capacity in the soil, causing water stress in plants. Erosion may reduce the depth of the root zone, the water-holding capacity of the remaining topsoil, or both.

The primary effect of erosion is loss of topsoil available for agricultural production. This type of damage occurs when the rate of soil erosion is greater than the rate of topsoil formation in any given parcel of land. Generally, the topsoil is the most workable and permeable soil, providing a hospitable medium for root growth, water retention and nutrient storage. Losses in topsoil affect these vital functions. Above and beyond the simple removal of topsoil, erosion leads to changes in soil structure, usually for the worse, causing problems of surface crusting, cultivation difficulties and uneven crop growth. The combination of loss of topsoil, tighter soil texture and exposure of subsoil can result in lower infiltration and hence higher rates of runoff.

One of the most insidious characteristics of erosion is the difficulty of detecting it by eyesight. Erosion often reduces productivity so slowly that the change may not be recognised until the land is no longer economically suitable for growing crops.

The concomitant removal of plant nutrients is also part of the erosion process. Subsoils generally contain fewer plant nutrients than topsoil, so sediments that runoff often contain more nutrients. This has often been expressed by saying that the erosion process is "selective" in that erosion removes the finer and more fertile particles, leaving behind coarse and infertile sands. Additionally, the impact of raindrops on surface can effectively seal the soil, rendering it impervious to water.

Erosion also affects soil structure by creating areas of surface seal, or crusting -- this nonuniformity makes farming more difficult. In particular, cultivation becomes difficult and crop maturity is often spotty. Soil texture changes, in that coarse grains of soil are left behind while finer grains are picked up and transported away.

Soil Erosion as an Economic Problem

Soil conservation projects are not cost-free. They require resources that could be allocated to other economic activities. Planners have to balance the current costs of soil conservation practices with the future values resulting from those practices. In assessing these tradeoffs there are several complicating factors. First, most of the off-site effects are not considered in the farmers' decision-making process and the amount of erosion produced by one rational individual may be much greater than what society would find optimal. Second, planning occurs under extreme uncertainty with regard to the value of agricultural productivity in the future, coupled with a high probability of irreversible damages. And third, processes to combat soil erosion have long-term returns that farmers may regard as far beyond their planning horizon.

Other Forms of Land Degradation[5]

Erosion is only one form of land degradation. Land is also degraded by accumulations of salts and alkali, organic wastes, infectious organisms, industrial inorganic wastes, pesticides, radioactive wastes, heavy metals, fertilisers and detergents. Saline and alkali soil problems are usually associated with irrigated lands. Organic wastes are often associated with the disposal of industrial, domestic and municipal wastes are the addition of salts and the specific ion effects of boron. Degradation due to infectious organisms is associated with disease and insects. Industrial wastes affect soils unfavourably because they release stack gasses (for example, sulfur dioxide, fluorides) with inorganic residues. Pesticides are very persistent in the soil and accumulations of chlorinated hydrocarbons have become a major concern. Radioactive effects on soils, though not now a major problem in LDCs, may become an important source of land degradation in the future. Traces of heavy metals have showed up in many foods because of dumping of industrial wastes. Although fertilisers are a source of soil nutrients, they are also associated with contaminants -- radioactive elements, for example, have been detected in raw rock phosphates.

Desertification and waterlogging are also important causes of land degradation. This book refers to them only in passing. However, the analytic frameworks that are explored in this book will also be applicable to projects aimed at controlling these forms of soils degradation.

Soil Conservation Methods

To remove or alleviate the major causes of soil erosion, policy makers may design and implement investment programs, or make institutional changes, or do both. Most investment programs include methods that may be classified into two groups -- engineering or mechanical protection methods and biological protection methods. Although this classification is rather arbitrary, it helps to identify the sources of benefits and costs of soil conservation projects. Without such methods, the costs incurred -- in the form of lost agricultural production, for example -- can be formidable.

As pointed out by Hudson (1981), many *mechanical* methods are used to control erosion. Stormwater diversion drains in ditches intercept stormwater that would otherwise flow down from higher ground onto arable land. Channel terraces, a form of earthwork at right angles to steep slopes, intercept surface runoff. Artificial watercourses, which include grass waterways, sod waterways, or meadow strips, can be used to drain runoff into water channels. Bench terraces, which convert a steep slope into a series of steps having horizontal (or nearly horizontal) ledges and vertical (or almost vertical) walls between the ledges can be constructed on the contour to minimize runoff, or with a slight radiant like graded terraces. Irrigation terraces, planted in raised strips between irrigation channels, are extensively used to grow rice, tea, fruit trees and other high-value crops. Orchard terraces are small level or reverse-slope terraces, each carrying one line of trees. Contour bunds, which are low banks thrown up by hand-digging approximately on the contour, are designed to conserve both soil and water. Pasture furrows are small shallow drains that spread out surface water as evenly as possible by not allowing it to concentrate in depressions or across watercourses. Their primary purpose is to conserve water by increasing infiltration, a technique often used on grasslands. Tied ridging consists of covering a whole surface with closely spaced ridges in two directions at right angles so that the ground becomes a series of rectangular depressions. Contour cultivation and grass strips involve tillage cultivation on contours between strips of grass or other close-growing vegetation. One last method is ridge and furrow, where the ground is tilled into wide parallel ridges interspersed with water-saving furrows. These combine erosion control with surface drainage (Hudson, pp. 92-107). Other mechanical methods are used for specific types of erosion. For controlling gully erosion, one can use wire bolsters, netting dams, brushwood dams, log dams, buckweirs, silt trap dams and the like.

Most of the *biological* methods of soil conservation involve crop rotation, pasture use, forest planting and management, strip cropping, planting of wind breaks and sand dune stabilization. One can classify different land use groups by the relative efficiencies of crop cover that protect the soil from erosion. A classification of this kind must be region and soil-type specific. In some regions, for example, permanent vegetation like trees may be more effective in protecting existing soil than certain row crops. With regard to cropping practices, different forms of tillage, planting methods, fertilizer use and harvesting methods greatly affect the productivity of soils.

The distinction between mechanical and biological forms of soil conservation is not just a matter of technique. They conserve the soil in essentially different ways. Mechanical methods *control* erosion -- that is, after the soil has started moving, prevent it from moving off-site. Biological methods prevent erosion -- that is, they intercept raindrops, thus not allowing the erosion process to start. Prevention is better than cure and certainly better than control.

Any of these methods generate both costs and benefits for the farmer. The scientist can help to assess the actual benefits of each method in saving soil. The task of the economist is to quantify in monetary terms the value of the potential net incremental benefits from these soil conservation practices.

Natural Erosion[6]

Erosion has always taken place and always will. The surface of the earth is constantly changing under the forces of nature and erosion is one aspect of this constant process of change. It is the starting point for the formation of sedimentary rocks and alluvial soils. The rate of geological erosion depends on climatic, topographic, geologic and other physical parameters. Hence, it varies greatly across the world.

Human related erosion is sometimes defined as geological erosion accelerated by human activities, but it is usually difficult to distinguish the two. Only in areas completely unsettled by humans can one be certain that humans have had no influence.

There are many places where the soil is particularly liable to such forms of erosion as landslides -- for example, the unstable mountain slopes of the Andes and Himalayas. However, it is not possible to say whether the incidence of landslides has been increased by population settlement or changes in land use. Another example is gully erosion, which may either be a straight geological process or be triggered or speeded up by man.

Humans are active agents of change and actions by humans usually accelerate the soil erosion process. However, there is a background erosion rate, known as "normal" or "geological" erosion, which is responsible for the sculpting of landforms (see Bennett, 1939) and continues unnoticed.

There are many forms of geological erosion -- leaching, oxidation, landslides and surface erosion. The speed at which geological erosion takes places varies from place to place. In desert areas, for example, the wind causes high rates of erosion.

The rate of natural erosion under undisturbed natural vegetation will be approximately the rate of soil formation, thus providing an overall soil balance. Several studies have been made in the attempt to quantify natural erosion. For example, the many experiments measuring the small amounts of erosion from areas completely covered by grass would suggest that natural erosion may be on the order of two tonnes/ha/year or less, depending on environmental conditions such as slope and rainfall intensity. In Zimbabwe, actual rates of soil removal have been measured on granite geologies at well below 1 tonne/ha. Certainly it may be said that geological erosion is very low.

Erosion therefore exists without human intervention in all parts of the world. However, the scale of the difference between human-induced and natural erosion can best be appreciated by referring to a classic experiment conducted in Africa (Hudson, 1981, gives further details).

Over a period of 10 years a bare soil site lost an average of 127 tonnes/ha, while a fully protected adjacent site lost less than 1 ton/ha/year. Differences in the amount of water runoff (and hence soil moisture to sustain plant growth) were no less spectacular. The difference between the sort of erosion focused on in this book and geological erosion can be at least 100:1, a ratio which emphasises just how critical the task of soil conservation is.

Inventory of Major Soils of the World[7]

Definitions

This section presents some information about the types of soil in the world. Particular emphasis is given to the risks of degradation.

The reliability of soil maps of the world differs from one area to another, depending on the quality and quantity of data. It appears that only one-fifth of the world's soils have been surveyed, with the highest percentage in Europe (76.3%) and the lowest in Africa (7.5%). The following is a standard classification of soils:

Ferrasols (Oxisols). Constraints result from low level of plant nutrients, low CEC and weak retention of bases applied as fertilizer. Sulphur deficiencies. High nitrogen losses, acidity, potentially toxic levels of aluminium. Very low calcium levels, resulting in limited rooting volume and moisture stress. A unit quantity of erosion is relatively very serious, the degree of seriousness depends on type of ferrasol. See Van Wambeke (1981) for a comprehensive discussion.

Acrisols (some Ultisols). Low nutrient levels. Trace elements deficiency. High nitrogen loss. Nutrients concentrated in surface horizon -- organic cycling essential. High erosion hazard and poor, variable structure. Surface waterlogging -- high runoff. Easily damaged by compaction. Very susceptible to declining productivity. Needs sensitive management.

Nitosols (some Alfisols, Paleustults, Paleudalfs, Paleustalfs). As for Ferrasols, but less acute nutrient deficiency. Argillic B horizon and texture change favour moisture retention.

Luvisols (some Alfisols). Moderate to high base status, but low in weatherable minerals. Deficient in nutrients and major elements. Moisture stress. Sensitivity to erosion. Low aggregate stability. Crusting. Sub-surface plinthite and hard pans. Very susceptible to continuing decline in productivity.

Vertisols (Vertisols). Physical, textural problems. High base saturation. Erosion and waterlogging problematic, but decline in productivity reversible with good management.

Planosols (Albaquults, Albqualfs). Low CEC and nutrients in topsoil. Waterlogging in wet season. Strong leaching and water stress in dry season. Erosion and decline in productivity variable.

Arenosols (some Psamments). Sandy, ferrallitic. Low in nutrients -- strong leaching. Poor response to fertilizer. Low field capacity -- drought stress. Loss in productivity very rapid.

Andosols (Andepts). Volcanic ash soils -- high water-holding capacity and natural fertility. High erosion but low decline in productivity.

Podzols (Spodosols). Acute nutrient deficiency. Coarse texture. High leaching and lack of nutrients. Very low retention of fertilisers and low available water capacity. Very low agricultural value in tropics. Highly susceptible to erosion. Rapid decline in productivity.

Cambisols (Tropepts). Brown earth without argillic horizon. Variable but with less pronounced limitations than soils with which associated (vertisols, ferrasols, acrisols mainly). Erosion often serious and productivity decline variable.

Xerosols, Yermosols, Solonchaks (Aridisols, Salorthids). Severe moisture stress. Salinity problems. Low natural productivity. Reclamation and irrigation needed.

Fluvisols, Glevsols, Histolsols (Fluvents, Aquepts, Histosols). Range of localised problems -- flooding, nutrient deficiency, acidity.

The above is a very generalised review of the productivity characteristics of tropical soils -- much depends upon specific local circumstances, levels of management, inputs, crop types and the like. Some of the major effects on the soil of erosion-induced loss in productivity include a change in water-holding properties of the soil and also a change in nutrients.

Geographical Distribution

Table 2.1 demonstrates the world distribution of soils based on Dudal's paper. Regosols and Arenosols constitute 10.1%, while Lithosols, Rendzinas and Rankers constitute 17.2% of world land area. Most soil characteristics relate to productivity -- they include nutrient status, drought susceptibility, permeability, erodibility, crusting and organic matter levels. Ferrasols, Acrisols, Nitosols, Podzols, Podzoluvisols and Andosols are subject to mineral stress. These soils cover approximately 22.5% of the world land area. Yermosols, Xerosols, Kastanozmes, Solonchaks, Solonetz, Regosols and Arenosols are subject to droughts. These soils cover approximately 28% of the world's land area.

Fluvisols, Gleyols, Vertisols, Planosols and Histosols are subject to stress in excess water. These soils cover 12% of the world's land area. Lithosols, Rendzinas, Rankers and Cambisols are relatively shallow. These soils cover nearly 24% of the world's land area. Chernozems, Greyzems, Phaeozems and Luvisols are soils in which stress features are least pronounced, account for approximately 10.1% of the world land area. Areas which are affected by freeze stress and permafrost cover 14.8% of the world's land area.

Table 2.1 Distribution of the Major Soils of the World

Soil Associations Dominated by	In 1,000 ha	In %
Fluvisol	316,450	2.4
Gleysols	622,670	4.73
Regosols and Arenosols	1,330,400	10.10
Andosols	100,640	0.76
Vertisols	311,460	2.36
Solonchaks and Solonetz	268,010	2.03
Yermosols	1,175,980	8.93
Xerosols and Kastanozmes	895,550	6.79
Chernozems, Greyzems and	407,760	3.08
Phaeozems	924,870	7.02
Cambisols	922,360	7.00
Luvisols	264,120	2.00
Podzoluvisols	477,700	3.63
Podzols	119,890	0.91
Planosols	1,1049,890	7.97
Acrisols and Nitosols	1,068,450	8.11
Ferrasols	2.263,760	17.17
Lithosols, Rendzinas and Rankers		
Histosols	240,200	1.82
Miscellaneous land units (icefields, salt flats, rock debris, shifting sand and the like).	420,230	3.19
WORLD LAND AREA	**13,180,390**	**100.00%**

Source: Dudal, R. "Inventory of the Major Soils of the World, with Special Reference to Mineral Stress Hazards". Food and Agriculture Organisation of the United Nations: Rome. (Unpublished) 1984. Reprinted by permission.

Notes

1. Analysis of soil deposition is particularly important in appraising sedimentation control programs.

2. It should be noted that it is not deforestation or logging *per se* that increase soil erosion; rather, it is inappropriate land use practices after deforestation, as outlined in Hamilton and King (1983) and Hamilton (1985).

3. Tillage affects not only the structure of the soil but also its

moisture content. Machinery developed for the more robust soils of temperate areas may cause damage when used on tropical soils.

4. It has been estimated that approximately 23% of the world's land is used for grazing livestock. Overgrazing is due to poor land management, uneven distribution of animals, and excess numbers of livestock in proportion to the carrying capacity of the land.

5. This section draws heavily on Rauschkold (1971).

6. Non-human related erosion.

7. The material in this section was derived from R. Dudal, "Inventory of the Major Soils of the World." FAO, Rome. Undated, unpublished draft paper and from Stocking (1984a), pp. 24-26. The soil types listed in the text represent only a subset, thus, Table 2.3's typology does not necessarily match the referenced list.

3

Soil Erosion Effects: A Typology

To facilitate the identification of benefits and costs resulting from activities to control erosion and to organize the empirical evidence on the magnitude of erosion in developing countries, this chapter defines the major erosion effects and develops a typology for the identification of such effects.

There are three types of soil erosion effects, "upstream effects," "downstream effects," and "worldwide effects." Upstream effects are the effects of erosion within a project area or in the upper portion of a river basin (for example, watershed). Downstream effects are the effects of erosion outside a project area but within the same watershed and usually within the same country. Worldwide effects are effects across countries or regions of the world. One can further classify upstream effects into, "on-farm or intra-farm effects" and "inter-farm or intra-area effects."[1]

Upstream Effects of Erosion

On-farm effects are the type of upstream effects most often recognized in the literature, though they are in many ways very difficult to quantify. They are chiefly losses in topsoil and corresponding decreases in land productivity. As stated earlier, these losses in productivity are due to losses of important plant nutrients, soil depth (root zone), water moisture capacity and other factors. If such losses go beyond a critical limit, irreversible soil damage and a permanent loss in land productivity may occur.

A large number of upstream entities can be influenced by erosion. Approaches to bridges, bridge footings and mainstream bridge supports may be washed away and foundations may be undermined by water action or landslides. This can result in difficulties in such activities as transport to and from a region. Erosion may also lead to siltation that affects irrigation infrastructure (clogging canals and tunnels) and ground-water irrigation equipment. Pump life, for example, may be shortened by abrasive action. Nutrients washing into the water may cause aquatic weeds to become a problem and the nutrient-chemical composition of water may render it less (or possibly more) suitable for irrigation and other uses. Finally, erosion may accelerate on-site flooding, with corresponding losses in human lives, animals and building.

Inter-farm effects also occur when actions by a farmer or a group of farmers affect the activities of other farmers. Examples of this are landslides which destroy another farmer's crops, clogging of irrigation canals, or even damage to existing infrastructure on other farms (for example, terraces).

Downstream Effects of Erosion

The erosion process also includes the transport of soil particles from one part of a drainage system to another. The transport process occurs via water or wind.[2]

The deposit of sediments may have both positive and negative effects. Because sediments often have a high nutrient content, they have the potentially beneficial effect of increasing land productivity when deposited on other agricultural lands. Many lands could never have been cultivated if they had not received sediment deposits from another part of the system.

However, the negative effects of sedimentation are our main concern here, since soil erosion projects are often designed to control the soil transport process. Excess sediments in rivers change the physical conformation of river basins and produce stream bank erosion and flooding of productive land during periods of high water flows. Some lands are thus irreversibly lost for agricultural production. Capital investments in houses, other infrastructures and livestock, as well as human lives, may also be lost.

An important negative effect of sedimentation is the silting up of water reservoir structures that are very expensive and often non-replaceable. These structures are non-replaceable in the sense that, for example, dam sites are not always available to build another dam. And if they are available, the construction of a new dam may cost a great deal more.

A dam acts as a sediment trap. The location of the dam and the magnitude of river flows in relation to trapping efficiency, will determine the speed at which the area above the dam will be filled.

Some degree of sedimentation above dams will always occur. However, when the rate of sedimentation is higher than was estimated during construction, it can have two major negative effects. It reduces the amount of water that can be stored above the dam and it may affect water quality negatively. Technologies are available to get rid of or wash away sediments, but if the rate of sedimentation is very high, a net loss in storage capacity will occur. This loss will harm the system's ability of the dam to curtail floods and its capacity to supply water for downstream irrigation or drinking water.

Because the soils that are washed away often contain a high concentration of chemicals, they change the quality of water. Excess nutrients in the water will stimulate the growth of weeds in a reservoir and in turn affect its storage capacity. Also, an excess of chemicals (for example, from pesticides) will constrain the availability of water for drinking purposes or result in higher costs for water treatment. Finally, chemical and hard sediments will cause physical or chemical corrosion of electric-generating facilities at dams.

Other effects of sedimentary deposits are,

(1) over-bank deposition -- burial of fertile soils by infertile sediments,
(2) damage to growing crops, whether by direct damage to the crop or indirectly by disrupting farm irrigation or drainage system,
(3) swamping -- impairment of drainage, with an accompanying rise of the water table and an increase in swampy areas of alluvial land,
(4) channel filling -- aggradation of natural or constructed channels, causing more frequent flooding, higher flood heights and interruption of irrigation and drainage systems -- channel filling may also change the channel course,
(5) reservoir deposition -- capacity loss in reservoirs caused by swamping above the delta of a reservoir and a rise in water tables due to impedance of groundwater inflow,
(6) other nonagricultural and agricultural effects,
(7) suspended sediment -- water supplies contaminated and made turbid by sediment that remains suspended in the water,
(8) recreational and environmental prolonged periods of excessive turbidity have damaging effects on fish, wildlife and natural vegetation.

Even low levels of turbidity reduce aesthetic values and may cause abandonment of recreational facilities. Swampy areas caused by sediment deposition provide breeding areas for mosquitoes and other insects.

Deposition in channels, harbours, bays and estuaries interferes with navigation and with marine life breeding areas and habitat.

Sediment yields are determined by, the physical features of the watershed, such as size of drainage area, the weather, the type of soil in the watershed, the cultivation practices (for example, fires, overgrazing, wood cutting), and the the structure of downstream reservoirs (trapping efficiency). The major causes of high sedimentation rates in developing countries are:

1. Changes that take place in the watershed after a dam is built. People who move into the watershed, may destroy the vegetation and thus cause more erosion and sedimentation.
2. Lack of planning to control the kind of land use practiced in the watershed.
3. Underestimates of sedimentation rate, due to lack of adequate data and records.
4. Failure to develop, or to implement programs designed to reduce erosion through different cultivation practices.
5. Problems in selection of dam sites. A small capacity reservoir serving a large watershed and having a large volume run-off is sure to have a shorter life.
6. Improper management of water in dams, resulting in inadequate removal of silt.

The costs of high sedimentation rates are often measured by the economic value of the output foregone due to changes in water supplies. However, the social costs may be much higher. When fossil fuel energy prices increase, either in the short or long term, hydroelectric power becomes a feasible alternative energy source for certain countries. But this comparative advantage is lost if sedimentation is not under control. Moreover, as stated earlier, dam sites are irreplaceable assets, the number of dam sites that would enable a country to exercise its comparative advantage by producing hydroelectric power is limited. As a country loses these sites due to high rates of sedimentation, future energy options are severely limited.

World Effects of Erosion

Transport of eroded materials does not occur solely within a region or a country. Recently published research shows that transport occurs across countries and even across continents. With the use of satellite imagery and other techniques, researchers have found regularity in the transport of dust across the oceans, Prospero *et al.* (1981)

Institutional Implications

The typology outlined here will not only help in identifying the benefits and costs of soil conservation projects, but will also allow planners to design effective institutional arrangements.

The "unit of account" of this typology -- aside from the worldwide effects -- is the total catchment area. It will be difficult for country managers to implement projects if the whole catchment is not seen as one unit. In addition if a sectoral view is adopted (for example, agriculture, forestry, water supply, hydro-power), investments, policies, and national standards will result in fragmented interventions.

To control erosion at the farm level, for example, the government will have to adopt arrangements which will change the behaviour of farmers. Local-level organizations involving farmers' participation should be adopted. On the other hand, if the most significant effects are downstream, institutional arrangements will have to take into consideration both up-stream and downstream economic agents.

Despite the absence of adequate methods for the monetary valuation of nonmarket commodities, this typology is also important for two other reasons. First, it is important for project design. In particular, assuming that a policy decision is made to use a least-cost approach rather than a benefit-cost approach to appraise a soil conservation project, one would require an adequate accounting framework. This framework should enable planners to properly identify the type of conservation activity, the necessary infrastructure and the needed project (or program) components to control erosion and sedimentation. Major negative trade-offs will occur if improper accounting of effects results during identification and design of soil conservation programs. An example of this problem is the construction of dams without consideration of a water shed management program.

Second the typology is important for setting policies and standards. To ignore either upstream or downstream effects will result in costly policy decisions. One example is where the erosion rate (per hectare) is considered to be relatively low but the resulting sedimentation rate is very high and causes major downstream damages. Another example is that of low erosion rates, with a very high chemical concentration in sediment particles. This will cause pollution of downstream waters.

The typology outlined in this chapter is both an instrument for identifying the benefits and costs of land conservation programs and a framework for designing effective institutions.

Notes

1. The typology of major effects from soil erosion is designed to facilitate analysis. It should be noted that the examples of upstream, downstream, and worldwide effects are merely illustrative, and that analysis should focus on what is happening and where. For example, an up-stream project area could well have a reservoir in it; therefore, effects on the reservoir would be counted as upstream effects. Another typology divides soil erosion effects into two categories--on-site and off-site. In this case, the project site is the determining boundary, and any and all effects that occur within the site would be included in the initial analysis. Of course, many important effects occur off-site (e.g., sediment deposition, flooding, changes in water quality), and a socioeconomic analysis should include these effects as well. The on-site typology is thus also a useful way to organize empirical evidence (Dixon and Easter, 1984; Dixon in Hamilton and Swedaker, 1984). For either typology the key is careful identification of all major effects, whether they are easily valued or not.

2. The terms "sediment" and "silt" are frequently used interchangeably to describe transported soil particles. Strictly speaking, *silt* is a type of soil particle intermediate in size between sand and clay, while *sediment* is any matter that settles out in a liquid. Therefore, silt is transported by water and is deposited as a sediment. The deposit of any type of soil particle is frequently called both sedimentation and siltation.

4

Extent of Soil Erosion in Developing Countries

Introduction

The extent of soil erosion, particularly in developing countries, is such that serious constraints to economic growth will occur in the medium and long term. The main consequences will not only be seen in losses in agricultural productivity but also in substantial foregone economic benefits in other sectors of the economy.

Land is a relatively fixed asset. The amount of land available is limited and the initial quality of soils is given. Soil quality greatly influences a country's future development. When rapid erosion takes place, irreversible damage may occur, such as a constant diminution of the overall area of arable lands. However, the regeneration of land assets is very slow. "It takes anywhere between 100 to 400 years for one centimetre of topsoil to be formed in nature while much soil can be lost in just one year due to erosion" (Library of Congress, 1979, page 15). When erosion reaches very high levels it poses a threat to the ability of a country to produce food supplies and export crops. This translates into substantial losses in human well-being.

The main objective of this chapter is to assess the extent of erosion in developing countries. Although the data are still sparse, statistics at the national and local levels are sufficient to show that this problem should be a major concern at all levels of decision making.

After providing background information, this chapter presents the existing empirical evidence on erosion at the country level, the extent of erosion effects at the farm level, the impact of erosion on downstream economic activities (including estimates of sediment loads of major rivers), sedimentation rates at major dams of the world, flood damages and water quality changes and global estimates of erosion rates in developing countries.

The reader should use caution in interpreting the data presented here. Given the limitations of the data base (for example, the fact that the data were compiled from many different sources), one risks making incorrect comparisons or generalizations. In addition, the data presented here come from direct estimates, often carried out by experiment stations near a farmer's fields, or at different points along a river basin. Other data come from estimates obtained using simulation models. Such models have a large number of shortcomings, particularly when applied to areas with a climate and environment other than that from which the model was derived. To make rough comparisons between data collected at the farm level and data on sediment loads in rivers, a common unit of measurement is used, tonnes of topsoil per hectare per year.

Land Use Patterns[1]

Differences in soil and climate have produced an almost infinite variety of cropping systems in developing countries. Five main crop zones can be identified however, on the basis of the staple food crop that predominates in each.

1. **Rice:** which first grew on water-retentive soils in the humid tropics of Asia, has been adapted to a wide range of environments. Farmers now grow rice in the river valleys and coastal plains of south China, South and Southeast Asia, the Indonesian and Philippine Islands, Japan and Korea, as well as small areas of Latin America and East and West Africa. In many areas with more permeable soils and high rainfalls, upland rice is grown with other crops.
2. **Starchy root crops (cassava, yams):** are grown in areas of the humid tropics where soils are less fertile and not well suited for cereal cultivation, such as Western and Central Africa and parts of Oceania and Latin America. Cassava has also spread to northern Thailand, where it has emerged as an important crop.
3. **Maize:** is the most important staple in the subhumid tropics of Latin America and Africa. The most common crops farmed with maize are cotton, groundnuts, soybeans and sorghum in the drier areas -- coffee, cocoa and starchy root crops in the wetter areas.
4. **Sorghum:** is the main food grain in the wetter parts of the semi-arid tropics, while millet is the main food grain in drier regions. Groundnuts, cotton, cowpeas and pigeon peas are the most common associated crops.

5. **Wheat:** is the most important grain in much of the temperate zone and is grown in an increasing area of the cooler tropics as a winter crop in association with monsoon-grown grains or cotton.

For centuries, farmers increased their output mainly by increasing the amount of land they farmed. This is no longer the cae. In the past two decades, increased acreage has accounted for less than one-fifth of the growth in agricultural production in developing countries and for an even smaller fraction in developed countries. Nonetheless, there is still a great deal of unused arable land -- estimates for the developing countries range from 500 million to 1.4 billion hectares, compared with about 820 million hectares currently under cultivation.

Most expansion of farmland takes place spontaneously as farmers make use of grazing areas and areas previously forested. Farmers are also switching to permanent cultivation, especially in Africa and are reducing fallow periods. In the early stages of migration, farmers moved to the most productive land. Later, as population pressures force them into more marginal areas, their arrival causes erosion and a decline in soil fertility. Deforestation is a particular problem. Between 1900 and 1965, about half of the forest area in developing countries was cleared for agriculture. Although forests still cover half the land in the humid and semi-humid tropics, forest cover has been reduced to 15 to 10 percent in the semiarid tropics and the temperate zone. Massive deforestation has highlighted the value of forests. They regulate the speed at which rainfall runs off, prevent soil erosion, replenish nutrients in the soil and influence the local climate.

Irrigation has made the largest contribution to increased agricultural production in much of Asia, North Africa and the Middle East. The world's irrigated area has grown by 2.2 percent a year since 1960. Some 160 million hectares, one-fifth of the harvested land in developing countries, is now irrigated. About 60 percent of all fertilizer is used on this land, which produces over 40 percent of all annual crops in the developing world. Between 50 and 60 percent of the increase in agricultural output in the past twenty years has come from new or rehabilitated irrigated areas. China (with 49 million irrigated hectares) and India (with 39 million) accounted for mor than half the developing world's irrigated area.

While irrigation has many advantages, the fact remains that rainfed areas comprise 80 percent of the developing world's cultivated land and support nearly two-thirds of its farmers. Yield increases still depend on subtle interactions among soil, water, seed and sunlight. Soil erosion and declining fertility are the main threats to rainfed agriculture in the humid and subhumid tropics. Tackling them requires protecting the soil by continuous crop coverage and minimum tillage, as well as by drilling seeds and controlling weeds. Highly acidic, infertile soil in some Latin American countries present a rather different challenge, however. There, research focuses on reclamation, new crop rotations and more effective means of fertilizing the soil.

Livestock farming in the arid natural grasslands of the developing world continues to face some intractable obstacles to growth. Animals need a great deal of land, but in these areas it is often of poor quality and ownership may be ill-defined.

In many African and Asian countries, such traditional resources as wood, fuel, water, grazing land, herbs, medicines and crop land are common property. Improving the quality of common property lands would be in the interest of all, but it is not in any individual's interest to invest in them unless certain conditions are met. As a result, improvements are seldom made, the land is often alienated to other uses, remaining land is heavily over-utilized and excessive erosion is the inevitable result.

Rainfall Patterns[2]

Rainfall is an important determinant of erosion. Raindrops that splash on the surface detach soil particles at a rate that depends on the type of soil, vegetation cover, topography and other climatic and environmental characteristics.

Areas of the world that are subject to rainfall of higher intensity and more concentrated geographical distribution are potentially subject to higher erosion rates. Rainfall patterns play a crucial role in the erosion process. Other things being equal, a climate with rains concentrated in a short rainy reason (for example, seasonal wet-and-dry tropics) has a higher potential for soil erosion than one where rain is more evenly distributed. A climate where rainfall ranges from 500 to 700 mm annually (semiarid to savanna) has maximum erosion potential -- enough rainfall for some intense storms each year but not enough to sustain a protective cover of vegetation throughout the year. A recognition of the importance of rainfall patterns is essential to understanding the erosion hazard in developing countries.[3]

Climate Regimes[4]

Climate determines to a large extent the type of soil and vegetation in a given region and thus influences the utilization of the land. Together with other factors, climate determines the ability of land to support a population.

Several criteria are used for climatic classifications -- temperature, precipitation, vegetation and soils, temperature and precipitation combined and air mass source regions and frontal zone.

Classifications by temperature include equatorial, tropical, sub-tropical, middle latitude, subarctic, arctic and polar climates, when precipitation is used, the classifications listed in Table 4.1 are made:

The Koppen system combines temperature and precipitation in five major categories as follows:

1. **Tropical climates.** Average temperature every month is more than 64.4°F (18°C). These climates have a no winter season. Annual rainfall is large and exceeds annual evaporation.
2. **Dry climates.** Potential evaporation exceeds precipitation on the average throughout the year. No water surplus, hence, no permanent streams originate in B climates.
3. **Warm temperate (mesothermal) climates.** Coldest month has an average temperature less than 64.4°F (18°C), but more than 26.6°F (-3C). The C climates thus have both a summer and a winter season.
4. **Snow (microthermal) climates.** Coldest month average temperature less than 26.6°F (-3°C). Average temperature of warmest month more than 50°F (10°C), that isotherm coinciding approximately with poleward limit of forest growth.
5. **Ice climates.** Average temperature of warmest month is less than 50°F (10°C). These climates have no true summer.

Table 4.1 Classification of Climate Regime

Climate Type	Rainfall Type	Annual Precipitation
Arid	Scanty	0 - 10 inches
Semi-arid	Light	10 - 20 inches
Sub-humid	Moderate	20 - 40 inches
Humid	Heavy	40 - 80 inches
Very wet	Very heavy	over 80 inches

Erosion at the Country Level

Land degradation is rather severe in many developing countries. It is estimated that 55 percent of the total land in India is now subject to different forms of land degradation. Estimates of the area subject to erosion vary from 140 to 150 million hectares, out of a total area of 328 million hectares. Of these, 90 million are affected by water erosion, 7 million by salinity and 20 million by floods. The soil which is lost carried with it large amounts of nutrients. It has been estimated that Indian farmers lose 2.5 million tonnes of nitrogen (N), 3.3 million tonnes of phosphorus (P) and 2.6 million tonnes of nitrogen (N), 3.3 million tonnes of phosphorus (P) and 2.6 million tonnes of potash (K) each year (see FAO 1981). These nutrients must therefore be supplied artificially. Their value at 1984 prices amounted to US$6 billion.

In Turkey, planners have estimated that 75 percent of the land is affected to some degree by erosion (see FAO, 1981). It has been estimated that 20 percent of the total land area has been moderately eroded, 37 percent severely eroded and 17 percent very severely eroded. The main causes of erosion are uncontrolled land clearing and overgrazing.

In Guatemala, 40 percent of the productive capacity of the land has been lost because of erosion. National reports state that at least 63 percent of the land is moderately to severely eroded (see Library of Congress, 1979). Due to topographic factors, susceptibility to water flows is very high. Thirty to 50 percent of the pre-existing forest resources have been destroyed since 1950, 35 percent during the last ten years alone. There are several areas in the country (for example, Cerro, Alaska) where farmers have completely abandoned the land because it has become uneconomic to carry out agricultural activities.

In Mexico, 90 percent of the land suffers from various degrees of soil erosion, while in Argentina, 13 percent of the total cultivated area, equivalent to 18.3 million hectares, if affected by water erosion and another 16 percent, equivalent to 22 million hectares, is affected by wind erosion.[5] About $20 million is spent every year to dredge sediments from rivers and harbours.

Sand covers 40 percent of Mauritania. In Haiti there is no top quality soil left so that the country loses 14 million m^3/year (see Das, 1977, page 26). In Nigeria the topsoil lost every year from each hectare carries with it between 7 to 19 kg of nitrogen. If one takes Central America as a whole, which comprises approximately 8,500 square miles of land, the losses in top soil are estimated to be 50 million tonnes a year.[6]

Table 4.2 Selected Estimates of Erosion Rates in a Sample of Developing Countries

Country or region	Area Eroded	Area Eroded %	Annual Rate of Erosion	Remarks	Source
Ethiopia Plateau Watersheds	150 km^2 14 km^2	n.a	2.3 mill t 0.2 mill t	152.7 t/ha 165 t/ha	El-Swaify
Lesotho	n.a.	n.a.	224 t/ha	n.a.	Turner
Niger Ibohaname Dam	n.a.	n.a.	31 t/ha	n.a.	AID
Tahoun Watershed	117km^2	n.a.	468,000t	40 t/ha	El-Swaify
Nigeria East Central	n.a.		13 mill t	n.a.	El-Swaify
Bangladesh	n.a.	n.a.	84 mill t	n.a.	FAO
India	140 mill ha 150 mill ha	43 51		n.a.	FAO,
Indonesia					
Citarum Watershed	n.a.	n.a.	3.1 mill t		El-Swaify
Cimanuk Watershed	n.a.	n.a.	25 mill t		FAO,
Solo Watershed	n.a.	n.a.	13.2 mill t		El-Swaify
Citandux	n.a.	n.a.	9.5 mill t		
Brantas Watershed	n.a.	n.a.	6.2 mill t		
Upper Solo	n.a.	n.a.	17.9 mill m^3		
Korea					
Ichecon, Gochong	n.a.	n.a.	40 t/ha		Sung-Hoon Kim, Dixon
Nepal	n.a.	n.a.	240 mill t	35-70 t/ha	USAID World Bank
Philippines	11 m ha	30	n.a.	cultivated land	El-Swaify

Table 4.2 (cont)

Argentina	40.3 mill ha	13	n.a.		FAO, 1981 Musro
Argentina & Bolivia La Plate River Basin	n.a.		95 mill t	1,880 ton/km^2	OAS, Cuenca de Rio Plata
Barbados, Scotland	n.a.	70	n.a.		El-Swaify
Boliva Central Valley	112583 ha	34	n.a.		Beattie
Ecuador High Sierra	n.a.	54	210-564		
Western Prarrie	n.a.	40	ton/ha		El-Swaify
El Salvador	n.a.	57-77	n.a.		AID
Guatamala	n.a.	57	5-35 t/ha		AID
Xaya-Pixcaya	n.a.	n.a.	801,000 mill t		
Haiti	n.a.		14 mill m^3		Riding
Honduras Ulva	n.a.	n.a.	55 mill m^3		AID
Choluteca	n.a.	n.a.	22 t/ha		
Jamaica Upper Yalluhs	n.a.	n.a.	90 t/ha		FAO
Mexico	34 mill ha	80	n.a.		Bartelli, El-Swaify

WFP (1984) stated that "in Ethiopia, in over one half of the country, more than 2,000 tonnes of soil per square kilometre are lost every year -- an area equivalent to one large enough to produce food for 12,000 families." In many instances the losses in topsoil are the direct result of the clearing of tropical forests. About 37 million hectares of tropical forest were destroyed in Africa between 1975 and 1980, as were 12.2 million hectares in Asia and 18.4 million hectares in Latin America.(Stocking 1984a).

It has been estimated that the soil loss in areas of Nepal that have been cleared for subsistence cultivation ranges from 35 to 75 tonnes per hectare per year. Erosion in gullied areas is 200 to 500 tonnes per hectare per year. It has been said that the most precious export crop of Nepal is its soil.

On-Farm Effects of Erosion

The magnitude of the on-farm effects of erosion depends on many factors. This section presents data on soil and water losses, losses in nutrients, losses in yields and the effectiveness of crop management practices in controlling erosion.

The data show that losses of topsoil in developing countries are often much higher than acceptable tolerance levels.[7] Losses in "barren soils" are sometimes 100 times larger than losses in soils with vegetation cover. The data also show the influence of slope gradient. In Africa, barren soils with a 23% slope lose more than five times the amount of topsoil of those with a 4.5% slope.

Losses in plant nutrients are directly related to soil and water losses. The water dissolves soluble materials and carries away the lighter and finer particles. The runoff is therefore higher in organic carbon, nitrogen and phosphorus than the parent soil. These nutrient losses are greatest when crops are first planted and soil surfaces are relatively bare. After a plant provides a canopy and takes up available nutrients, erosion carries away nutrients at a much lower rate.

One of the most telling consequences of erosion is the ratio of nutrients in eroded soil to nutrients in parent soil. Without exception, the removed soil is richer in nutrients. In certain cases the concentration of some nutrient elements in the eroded soil can be at least two times that in the original soil. This represents a direct loss of natural or applied nutrients.

All the above-mentioned factors affect *land productivity*. Table 4.3 shows the relationship between crop yields and topsoil depth and Table 4.4 shows how yields are affected by different ecological and farm management conditions. Very few reports present data relating global soil losses to yield effects in developing countries. For research findings on relationships between soil conditions, management and productivity, see Das, Mukherjee and Kaul (1980).

With regard to the effectiveness of *soil conservation practices*, several reports show the potential effects of these practices on yields and on better use of available nutrients. In the absence of constraints, erosion could seemingly be reduced to very low levels. However, the available data do not reflect the cost of reducing erosion (Stocking, 1984a).

Moreover, most alternative erosion control methods demonstrate diminishing marginal productivity. This is to say that after an optimal level of application, no significant change in productivity occurs. For example, mulches can be applied at several alternative levels with sharply diminishing productivity over the range reported (Meyer *et al* 1970).

Combinations of contour ploughing, terraces and grading will decrease erosion rates but at diminishing rates of return (Gupta and Babu, 1977).

The distinction between mechanical/engineering practices and biological practices is relevant here. For example, in the Phewa Tal Watershed (Nepal), the abandonment of overgrazed land reduced erosion by 58%, while terracing added another 15% reduction (Flemming, 1983).

Compared with bare soils, land protected by continuous natural forest cover shows almost no erosion. Under the right conditions, reafforestation of degraded areas can help to rehabilitate the land and bring erosion rates under control. However, the type of tree species has to be chosen carefully. In Colombia, for example, grass ground cover was shown to be more effective than rainforest in reducing low erosion rates. This may be true even where slopes under grass are greater.[8]

Slope is positively correlated to erosion in all cases. The effectiveness of control measures may vary with the slope factor. In Nigeria, intercropping is relatively more effective in controlling erosion from moderate slopes than from flat or steep slopes.

Summing up -- the data presented in this chapter are incomplete, but they do show that both environmental and social conditions in developing countries are conducive to very high erosion rates -- possibly several orders of magnitude greater than either a "tolerable" soil loss or the actual rate of recreation of soil. Such losses cannot continue and are being felt increasingly in the degradation of the environment and a spiralling loss in land productivity. Much more research is needed to quantify the precise effects of soil erosion.

Table 4.3 Relationship Between Topsoil and Corn Yield

	Corn Yield (bushels per acre)		
Topsoil Depth (in)	Range	Average	Decrease in Yield
0-2	25-56	36	38
2-4	28-69	47	27
4-6	39-83	56	18
6-8	49-97	65	9
8-10	50-102	69	5
10-12	50-125	74	-

Source of material in this table, Pimental *et al* (1982), pp. 149-155 and Hufschmidt *et al* (1981).

Table 4.4 Effects of Conservation Measures on Soil Loss

Type of Control	Location	Baseline Soil Loss[1]	Relative Soil Loss (%)	Notes[2.]
Plant density	Rhodesia Zimbabwe	4.4	5.8	Maize 24,500 plants/ha Maize 36,675 plants/ha Intensification possible with fertilizer[a.]
Terraces	Nepal	34.7	73	Overgrazed pasture[b.] Improved terraces vs.
Bunds		40.0	68	unimproved[c.]
Contour ploughing	Brazil	52.8	6.3	Up/down ploughing & sowing, sandy soil
		21.4	80	Up-down ploughing, clay soil[d.]
Ridging	Ghana	1.8	90	2% slope, July-October[e.]
Mixed cropping	Tanzania	136.9	47	Bulrush, millet, sorghum
	Mwapwa	119.3	39	Bulrush, millet sorghum vs. conventional cult
Mulch	Indonesia	500	97	Bare soil
		200	93	Conventional cultivation

[1] Baseline soil loss is in tonnes per hectare on comparable bare soil, unless noted.

[2] (a) Fournier (1967), pp. 54-96. (b) Flemming (1983), pp. 217-288.
(c) R.C.V.P. Economic Paper on Selected Hill Areas of Nepal. Prepared by SEGIO 1900.
(d) Lal, in Greenland, D.J. and R. Lal (eds) (1975), pp. 143-150.
(e) Bonsu (1080), pp 247-283. (f) El-Swaify *et al.* (1982).
(g) Suwardjo and Abujanin

Soil Erosion Problems in Asia[9]

Introduction

The loss of large volumes of topsoil from upland watersheds of Asia presents a grave threat to the livelihood of local residents and to downstream water users. In light of this FAO, sponsored a Government Consultation on Watershed Management for Asia and the Pacific which was held in Katmandu, Nepal, in December 1982. The objectives of the consultation were, to examine the problems of degradation of upland areas in Asia and the Pacific, to assess the needs and constraints, as well as the capabilities, of the countries of the region with regard to the implementation of watershed management programs, and to explore the possibilities for technological operation and joint efforts in watershed management. Data assembled for the Consultation indicated that many of the factors contributing to excessive soil loss and preventing the introduction of effective control programs are shared across countries. Similarly, the proposals made at the Consultation indicate the commonality of the types of action needed throughout the region.

The Asian continent covers a very large proportion of the earth's land area and has a population of nearly half the world's total. While culture, climate and topography vary widely across the region, a number of factors combine to make soil erosion a problem of common concern. Six of the major rivers of the world -- the Indus, Ganges, Brahapuatra, Yangtze, Yellow, Mekong -- are in Asia and soil loss in the upland regions of these watersheds has similar effects, including flooding, siltation of reservoirs and reduced life of hydroelectric installations. Furthermore, in a region where agriculture accounts for a large percentage of each country's GNP, the adverse effects of erosion on agricultural productivity cannot be dismissed.

Causes of Erosion

Shifting cultivation was the most frequently identified cause of land degradation in the region. In Bangladesh almost 1 million hectares of forest land have been degraded by shifting cultivation. In Thailand the area cleared by shifting cultivation annually has been put at 300-400 square kilometres and it is estimated that the total area affected by shifting cultivation is greater than 5,000 square kilometres. It has been estimated that the annual monetary loss to the Thai economy from losses in land productivity due to shifting cultivation amount to US$93 million. In Laos, shifting cultivation is estimated to degrade 80,000 ha/year.

Closely related to shifting cultivation is the problem of illegal encroachment on forests by subsistence farmers. Forced by land shortages to clear and cultivate forest land, settlers invade steep slopes that are generally unsuitable for sustained agriculture. Encroachments are often the result of wars or, in some cases, the inundation of land behind newly constructed dams. In addition to the fact that land settled by encroachment is seldom suited for sustained agriculture, the poverty and insecurity of tenure of the settlers discourages them from investing in appropriate soil conservation measures.

Other problems mentioned in various country studies were over-exploitation of common property land for fuelwood and building poles and the grazing of domestic animals. Less frequently mentioned factors leading to soil loss were the settlement of land for commercial agriculture, as in Thailand, the use of inappropriate forest harvesting techniques (that is, land clearing) in Indonesia and poor design of roads and infrastructure in the Philippines.

Apart from these common features, certain factors peculiar to individual countries were described that may lead to excessive soil loss. In Fiji, for example, sugarcane is grown under a quota system. For farms on sloping land, quotas are estimated on the basis of the area that can be ploughed and fertilized whose slope does not exceed 11 degrees. In practice the 11-degree slope limit is often ignored and cultivation on slopes up to 20 degrees is not uncommon.

In the Philippines, indiscriminate open pit or underground mining has been identified as a major cause of erosion. The proceedings note that "disrupted soil and shallow underlying bedrocks, usually associated with steep slopes, bare soils and poor drainage systems, further enhance soil detachment and degradation of the watershed and siltation of infrastructure and downstream development."

Erosion caused by open-pit mining in the Upper Agno River Basin threatens to shorten the life of two hydroelectric dams at Ambauklao and Binya.

Effects of Erosion

The effects of soil erosion are also similar across countries. For example, the Consultation noted that increased flooding and siltation of waterways in Bangladesh is being caused by soil lost from the mountain catchments of upstream countries.

The downstream effects mentioned most often were floods and increased frequency of floods caused by siltation of reservoirs and riverbeds. These were noted in Bangladesh, Pakistan, Thailand, Fiji and the Philippines.

In the Philippines, annual flood damages in excess of 100 million pesos per year were reported. In Western Samoa, severe flooding has occurred more frequently in recent years and is correlated with upland watershed degradation. In 1974 in Western Samoa a major bridge was destroyed by large logs carried by a flooded river, in February 1982 the Apia area was damaged by floods that caused damages estimated at US $300,000.

One concern mentioned in the reports on several island countries was the effect of sediment flows on marine ecosystems. For example, it is feared that sedimentation and turbidity in the Vaisigano and the Fulvasou rivers in Western Samoa may harm offshore shrimp fisheries.

Institutional and Organizational Issues

The Consultation also highlighted the common institutional and organizational problems faced by Asian countries. Just as erosion problems do not observe national boundaries, neither do they remain under the strict purview of any one organizational entity. Thus, responsibility for erosion management is shared in Asian countries by such agencies as the National Ministry of Agriculture, provincial or state ministries, hydroelectric power authorities, public works agencies and environmental quality agencies. Coordination of the activities of these agencies, such as the development of national plans as well as implementation of soil conservation projects is difficult. In response to this organizational problem and in order to gain a multidisciplinary approach to erosion control, the government of Thailand has established a Watershed Management Division under the Royal Forestry Department. This division was to coordinate activities in land surveying and planning, forest extension, watershed rehabilitation and development, forest engineering and research and experimentation. Lacking sufficient resources however, the Watershed Management Division has been unable to perform its mission effectively.

In Bangladesh, responsibility for soil conservation was divided among three divisions -- Soil Survey, Soil Resource Development and Technical Support Services -- all within the Department of Soil Survey, Ministry of Agriculture. According to a plan prepared with assistance of the FAO/UNDP, the activities of the three divisions could be consolidated under a Watershed Management and Soil Conservation Division. This division will have six sections -- Planning, Engineering, Cartography, Research, Extension and Administration and Accounts -- and will prepare and implement management plans, develop research, provide extension services and prepare policy and legislation. The division will have both headquarters staff responsible for policy formulation and field staff responsible for implementation.

Proposed Solutions

The types of interventions described by the Consultation can be separated into five broad groups:

1. *Planning exercises*, including the preparation of legislation and national policies and project preparation activities.
2. *Land settlement schemes,* which are more or less integrated rural development efforts intended to provide farmers with sustainable

alternatives to shifting cultivation and other ecologically destructive practices.

3. *Forestry investments,* including afforestation, reforestation, various social forestry schemes and improvements in natural forest management.
4. *Education and research,* including professional and in-service training of staff and establishment of research/teaching centers.
5. *Downstream engineering works,* such as dredging of harbours and reservoirs and reinforcement of dikes.

While all of these types of programs have been attempted in Asia, the primary approach has been through forestry projects. Typical projects including large-scale industrial plantations are underway in Pakistan, Tonga and Sri Lanka. More recently, efforts in social forestry, including village and farm woodlots in Nepal and India and shelterbelts in Burma, have received greater emphasis.

A program being implemented by the Forest Industries Organization (FIO) in Thailand to establish forest villages and to promote agro-forestry practices illustrates the range of activities that are included in social forestry programs. FIO villages are established in forest reserves with up to 100 families per village. Villagers are recruited from shifting cultivators and are provided with small amounts of land for homesteads and backyard gardens. Each family is also provided with 1.6-2.4 ha each year for cultivation of an interplanted mixture of trees and food crops. Interplanting is permitted for two years at which time an additional plot is assigned to each family. Each family is paid for tree planting labor and is allowed to keep its production of food crops.

Second to forestry projects are a variety of other soil conservation programs, including settlement schemes and multipurpose watershed development programs that are primarily aimed at assisting small upland farmers. These programs have the joint purpose of raising rural incomes and encouraging the adoption of soil conservation practices. For example, in Bangladesh, the Clittayong Hills Development Project is intended to assist 2,300 families established permanent settlements to raise both cash and subsistence crops. In Pakistan, the Tarbella Dam Watershed Management Project has worked with farmers to limit overgrazing while planting fruit trees and maintaining flood terraces.

Other programs underway throughout the region include the development of legislation and national plans for soil conservation and watershed management, professional and in-service training of staff and research and development activities such as using different tree species and evaluating soil conservation measures.

Conclusions

The discussion at the consultation suggested that at least four tasks must be confronted in order to reduce soil erosion damages in the region, first, the establishment by government of priority for soil erosion control and the development of a legal framework that will foster mobilisation of human and financial resources, second, the development of institutional capacity that can translate incremental resources into effective programs, third, the integration of national resource management with overall national development planning and implementation frameworks, leading to the preparation of multi-objective projects, and fourth, creation of efficient incentives to encourage the active participation of small farmers in upstream areas.

Erosion in China's Yellow River Basin

The Yellow River in the Peoples' Republic of China carries the highest suspended sediment load of any of the major rivers in the world. The river is 5,000 kms long and drains an area of 680,000 sq kms. Erosion in the river's watershed has required the diversion of resources to prevent flooding and has disrupted hydroelectric power development schemes. Terraces and other structures have been constructed in an attempt to control erosion.

The Yellow River flows eastward from its source in Tibet and northward through mountains, grasslands and deserts. The river then turns eastwards and south through Shansi and Shensi Provinces. In this region the river flows through plateaus which are extraordinary erodible. These areas have been overexploited by agricultural expansion, the harvesting of forests and continued destruction of the original grass cover. Here severe erosion occurs, delivering heavy sediment loads to the river. In Northern Shansi and Northwestern Shensi Provinces, soil losses range form 20 to 200 tonnes/ha per year and average 45 tonnes/ha per year. The river then flows east across the North China alluvial plain, emptying into the Pohai Sea.

In its middle reaches the river carries its highest sediment load of about 700 kg/m3 (43.7 lbs/sq.foot), or about 50% silt by weight. At a silt concentration of 56%, silt and water become plastic, meaning that the flow is essentially a liquid mud. Deposition of sediment on the river bottom occurs in the lower reaches of the river at a rate of 5 to 10 meters above the surrounding countryside. By the time it reaches the sea, the river's sediment concentration falls to 40 kg/m^3, which over the entire catchment represents soil losses of more than 1 centimetre (0.4 inch) per year. Erosion is more than 28 tonnes/hectare per year over the entire area.

Although the heavy sediment load of the river has hindered the construction of major dams, the growing population's need for power and water storage led in the 1950s to the construction of a major dam at Sanmerxia Gorge. Completed in 1960 with design and construction assistance from the USSR, the Sanmer dam is 98 meters high with a storage capacity of 33 billion cubic meters. After two years of operation it was discovered that the reservoir was silting up at a rate greater than originally expected. The original design assumed an annual sediment load of 1-4 billion tonnes, of which 40 percent would pass through the dam. It was also assumed that a 3 percent per year reduction in upstream erosion could be achieved. In fact, sediment loads were one-third greater, the dam's trap efficiency was 90 percent and no reductions in erosion were made.

In order to reduce sedimentation, the Chinese began sluicing water through the dam in 1962, in 1965 they opened diversion tunnels and converted penstocks to sluice ways. The original high-head turbines have been replaced by fewer low-head turbines, reducing power generating to one-third of planned capacity. Adjustments in operations of the dam have reduced its trap efficiency to 21%, although this still means the deposition of 400 million tonnes of sediment in the reservoir annually.[10]

Major responses to the erosion problem in the Yellow River basin include the construction of bench terraces, check dams and tree and grass planting. Bench terraces, mainly constructed by manual labor, are now estimated to cover more than 50% of some provinces.

Soil Erosion in Certain Mediterranean Countries[11]

Introduction

Loss of topsoil is a severe threat to the economics of the Mediterranean region. The three human activities most responsible for soil degradation are, expansion of agriculture into fragile areas, an increase in animal populations, and the cutting of forests for fuelwood.

Loss of topsoil is disrupting the economics of rural areas. As productivity declines, incomes fall and migration increases. In addition, erosion threatens dams and parts that contribute to output. Governments in the region have attempted a number of interventions and a variety of lessons have been learned.

The Dimensions of Soil Erosion in the Region

There are no figures on the extent of the erosion problem in the region as a whole. However, there are a number of estimates for individual countries that provide an indication of the extent of the problem. In Morocco, 17.5 million ha, or 40% of the country's total land area, is exposed to erosion. In Turkey, 42 million ha, or over half of the land, is subject to erosion damage. A study of 1.05 million ha of land in four catchments in Greece found that 21% had experienced moderate erosion and that 46% had experienced severe erosion. In Greece as a whole it is estimated that 30 million cubic meters of soil are washed away each year.

In Tunisia in 1973 the flooding of Mezerda demonstrated the volume of soil loss taking place in its watershed. During a flood of six days' duration that inundated 470 square kilometres, 75 million tonnes of sediment were deposited. This corresponded to the removal from upstream watersheds of at least 100 million tonnes of soil and a concentration of 100 grams of sediment per litre of river flow.

These levels of soil loss are a threat to agricultural production and downstream infrastructure. In Greece, annual damages from soil erosion are estimated to be US$5 million. In Morocco, an annual loss of reservoir capacity of 54 million m^3 is estimated. Siltation of the Moroccan port of Kenitra at a rate of 30 million tonnes per year necessitates continuous dredging. In Tunisia, drifting sand dunes have completely covered several cases and wind erosion threatens others that are important to the populations of other southern regions. In Turkey, water-caused erosion is creating critical problems of siltation of reservoirs, ports and other infrastructure. In Greece, damage to the productivity of the soil has led to the abandonment of land and is forcing rural populations to migrate.

Causes of Soil Erosion of the Region

Climatic and geological factors make the region particularly susceptible to erosion. Rainfall in portions of Morocco and Tunisia is concentrated in intense storms with a high capacity to detach soil particles. In other areas of those countries, dry season winds are a major source of erosion. Steep slopes throughout the region are also subject to erosion.

Far more important causes of soil loss however, are human activities, especially the expansion of agriculture and livestock and the removal of forest cover. Population growth is now 2.7% per year in Turkey and 2.3% per year in Tunisia. As a consequence there is increasing pressure on the natural resource base.

In Turkey, 10 million people (25% of the population) live in or around forested areas. The livestock population of Turkey includes 21 million goats, 37 million sheep and 18 million head of cattle. These livestock cause severe damage to forests and pasture land. In Morocco and Tunisia the large rural populations are being forced to cultivate steeper and steeper slopes.

Responses of Governments and Lessons Learned

Governments in the region are aware of the severity of the erosion problems in their countries and have initiated programs to protect soil and forest resources. In Morocco, pilot projects covering 330,000 hectares have been carried out to control erosion. These projects will be followed by larger scale efforts. In Tunisia, 911,732 hectares have received some sort of soil treatment and 120,000 hectares were reforested between 1949 and 1971. In Greece, 100 catchments covering 1.1 million hectares, or 23% of the watershed area of the country, have been established. In Turkey between 1957 and 1975, 900,000 hectares have received erosion control treatment.

Although the success of these soil conservation programs varies, some general conclusions can be reached:

1. Sociological factors are important aspects of the erosion problem of the region and must be taken into consideration in designing conservation programs. Tensions arise because of the differences between settled agriculture and nomadic pastoralism, private property and collective ownership and free enterprise and central planning. The participation of local people in soil erosion control programs must therefore be assured.
2. Control of erosion requires a reduction of the pressure to use marginal agricultural and grazing land. Land use planning based on analysis of the capabilities of the land should be carried out.
3. Reestablishment and expansion of forest cover and of managing forests for multiple uses, including wood production, soil protection, water yield and range uses, it vital.
4. Erosion on steep slopes be reduced and agricultural and livestock production increased, through engineering works such as terraces.
5. The implementation of soil conservation programs requires a clear expression of objectives, as well as legislation creating an organizational framework that can execute multi-disciplinary projects, including research, monitoring and evaluation.

Ecuador, Deforestation and Drought[12]

Deforestation

Although half of Ecuador is still forested, deforestation and its negative agricultural and demographic impacts affects all regions of the country. The most serious deforestation in Ecuador is occurring in the Oriente region, where settlements have been established along access roads to the petroleum fields. In the Oriente and the Costa regions, government policy has hastened deforestation by promoting the colonization of virgin lands and encouraging clearing of lands for "productive use." If the current pace of deforestation continues, Esmeraldas province will be denuded within 20 years. Parts of the Sierra historically have been bare because of the arid climate, the loose porous volcanic soil and the absence of soil conservation measures.

Human activities at higher elevations of the Sierra are steadily eliminating the remaining vegetation. Improper farming practices and the clearing of forests for firewood and charcoal production are causing serious soil erosion. The areas most acutely affected are Chimborazo, Cotopaxi and Loja provinces in the Sierra, the extreme western areas of Manabi and Esmeraldas in the Costa and the Cordillera Oriental.

Soil erosion has contributed to downstream flooding by causing sedimentation of rivers and silting of flood plains in inter-Andean valleys. Deforestation also triggers landslide activity by reducing soil cohesion and moisture retention and is a major factor in the desertification process. Presently, about 5,000 hectares a year are being transformed into desert.

Although reforestation projects have been initiated in the Oriente and Costa regions, many schemes are ill-conceived and incomplete due to a limited understanding of the forests' complex vegetation. Some projects have resulted in the reforestation of the lower portions of steep hillsides but have left the critical higher areas bare.

As a result, the landslide potential which the projects were designed to minimize remains unchanged. Eucalyptus and pine are the most common species used for re-forestation, primarily because of their rapid maturation. Mountain slopes surrounding Quito have also been extensively reforested with eucalyptus.

Drought

Localized droughts recur in the Sierra on a multi-year cyclical basis and usually affect grains, tubers and other subsistence crops. Poor cropping methods and limited irrigation works aggravate drought potential.

In addition, the Santa Elena peninsula west of Guayaquil and some areas of Manabi Province are perennially arid except when an anomalous rainy season renders these locations fertile.

Notes

1. Part of the material presented here has been quoted from Chapter 6 of the *World Development Report 1982* Washington, D.C: The World Bank, 1982.

2. The material presented here is based on Jaeger (1983), Mooley and Parthazarathy (1983), Ratcliffe (1983), Nicholson and Chernin (1983) Yan (1983), Nemec (1983), Van der Leeden (1975), and Strahler (1969).

3. See Jackson (1977).

4. The material in this section is taken from Strahler (1969).

5. See Government of Argentina, *Conservacion de los Suelos.* Law No. 22, 428.

6. See UNDP/FAO, 1977, page 39.

7. The term "tolerance level" (T) refers to a threshold in terms of a "maximum" acceptable loss in topsoil. In the United States this T level has been estimated at 5-10 tonnes/hectare/year, a level believed to be consistent with sustained soil productivity. Since there is no strong correlation between T values and soil productivity, however, it would be misleading to overemphasise differences between actual (or estimated) erosion and the US T value. Five tonnes/hectare/year could be too high when applied to shallow and fragile soils, or too low when applied to deep soils. This T value was determined by "expert opinion" and has limited scientific value. It should be noted that eroded soil is transported within the same watershed and does not necessarily end up in rivers.

8. See McGregor 1980, Hamilton and King 1983 and Hamilton 1985.

9. This report is based on FAO, "Proceedings of the Government Consultation in Watershed Management for Asia and the Pacific." FAO: RAS/81/053. Katmandu, Nepal 5-13 December 1982.

10. At, say, 1.8 tonnes per cubic meter, this represents more than 200 million m^3 of lost capacity annually, or less than 1% of original capacity.

11. This report is based on a special issue of *Homme, Terre et Eaux*, vol. 9, no. 30 (January-February 1979).

12. Office of Foreign Disaster Assistance, Agency for International Development. *Ecuador: A Country Profile.*

5

Extent of Downstream Damages

Major Effects of Sedimentation

Introduction

This chapter is divided into three major sections, effects resulting from sedimentation, effects resulting from floods and other downstream effects and quantification of a global erosion rate in developing countries. Empirical information to illustrate the extent of some downstream effects of soil erosion in developing countries is presented here.

It is important to note that the literature on sedimentation often separates the data into two categories, sediment loads in rivers (that is, where measurement takes place "midstream") and sediment deposition in reservoirs (that is, where measurement takes place "downstream"). To quantify the amount of topsoil that reaches rivers, streams, lakes, dams and estuaries, the earlier presentation focused on the source of sedimentation (that is, on detachment of soil). If one uses an estimate of sediments delivered, a sedimentation rate from farmers' fields could also have been quantified (that is, quantification "upstream").

Each of these categories presents only a partial view of the effects resulting from sedimentation. Furthermore, many other things are affected by sediments -- for example, fisheries, recreation, port facilities, riverine transport and wildlife. As explained above, sediments are not all deposited in the reservoir areas of major dams. Some of the sediments flow to estuaries and oceans and some are deposited on riverbeds (referred to as "bed load"). Large volumes of sediment change the morphology of riverbeds, making the whole hydrologic system susceptible to flooding. Depending on the volume and the quality of sediments, turbidity results. This changes the quality of the water, animal habitats, river navigation and recreation. Also, port infrastructure, many times located near rivers or estuaries has to be dredged constantly due to the amount of sediments that reach the oceans. In Argentina, for example, it was estimated that the country spends at least US$20 million a year just to keep its main port near the mouth of the La Plata River functioning at adequate levels.

Thus, to trace the impacts of any given level of sedimentation on each of these activities is a task that goes beyond the scope of the book.

Sediment Loads of Major Rivers

The collection of statistics on the sediment load of major river basins was an important part of this study. Since the data set is quite large, only some of the material is presented in the text. The rest can be found in the accompanying tables. To make the statistics comparable with those presented in Chapter 4, an estimate of soil losses on a per hectare basis is also presented.

The conversion from sediment load to erosion rate is often done by using a Sediment Delivery Ratio (SDR). In other words, a field erosion rate may be 100 tonnes per hectare, but only a small proportion is totally lost from the catchment. The specific values of the SDR are highly correlated with the size of the drainage area of a given river, the larger the drainage area the smaller the SDR value (Klingebiel, 1972, Curtis, 1976).[1]

The SDR however, is only an approximation, since many other factors help to determine sediment transportation from one area to another. Sediment volume is highly variable, even within the same drainage basin. Moreover, since the type of soil, among other things, will also influence the ability of water to carry these sediments (that is, rate in suspension vis-a-vis deposition at the bed) and this ability is independent of size of the drainage area, much variability is found in the data. Factors like climate and topography are also important. The amount of sediment load in rivers in tonnes of soil per hectare erosion rate equivalent exceeds preestablished threshold levels, often by very large amounts.

Ranking the rivers by size of drainage area is important, since with rare exceptions there is a high correlation between size of drainage area and population. Such major rivers as the Ganges, Yangtze, Brantas, Nile and Niger all support very large numbers of people and often their catchment areas go far beyond the boundaries of one country.

Perhaps the most important implication of this fact is that the management of sediments will require major international efforts. For example, it has been noted that 90% of the total drainage area of major rivers in Bangladesh is located in neighbouring countries (see US Library of Congress, 1980). The effectiveness of sediment management policies in Bangladesh will therefore depend mainly on the management policies adopted by other countries.

The information can also be organized by ranking river basins by estimated erosion rates (or their corresponding volume of sediments in relation to drainage area). In this case, it is interesting to note that there are rivers with very small drainage areas that rank very high in terms of erosion rates. This shows that sedimentation is a complex phenomenon which results from the interplay of many important variables.

Land Use Upstream and Sediment Loads

Studies in the United States have shown that the major source of sediment is sheet erosion in areas primarily devoted to agriculture (Klingebiel, 1972 and Curtis, 1976). One study analyzed erosion in 157 watersheds and determined that 73% of the total volume of sediments came from sheet erosion, along with 10% from gully erosion (Curtis, 1976). In Zimbabwe, in a catchment with major gullies and an assumed severe gully erosion problem, sheet erosion rates were four times higher than gully erosion rates (Stocking, 1978 p 42-46).

Weather is also an important factor. For example, reported peak sediment yields from a single severe storm, sedimentation was several times higher than under normal weather conditions. This is an important consideration, particularly in countries where rainfall distribution and intensity tend to be concentrated within a short period of time. Rapp (1977) reported that rates of sedimentation in the semiarid region of Tanzania fluctuated between 200 and 730 m^3/km^2/year depending on rainfall pattern and distribution.

Land uses other than farming also affect sedimentation rates. In Malaysia, four sampling stations were selected in the Sungei Gombak Catchment to study the major factors affecting sedimentation. Sedimentation was most significant at the station where such activities as tin mining were found. The sedimentation rate was high because of chemical weathering and destruction of vegetation.

In Honduras (Castellanos *et al.*, 1980), human encroachment in the Los Laureles watershed has been very significant. Estimates of the sedimentation rate are presented in Table 5.1. From this, one can immediately see how alternative forms of land use affect sedimentation rates: 2% of the land area is responsible for 45% of the sediments.

Changes in land use in general and in agricultural practices in particular, can play an important role in reducing sedimentation. Several studies have demonstrated that the cultivation of potatoes in Idaho results in very high volumes of sedimentation.(Klingebeil 1973).

Table 5.1 Sedimentation Rates at Los Laurales Watershed -- Honduras

Land Use	Area %	Sedimentation (%)
Roads and trails	2	45
Hill cultivation	13	20
Grazing land	20	20
Burned forest lands	10	4
Brush lands	20	1
Forests	35	1

Source: Castellanos, V, and J.L. Thomas. *Application of Multiple - Use Research on Watershed in Honduras*. Paper presented at the IUFRO/MAB Conference: Research on Multiple-Use of Forest Resources, May 20-23, 1980, Flagstaff, Arizona. Reprinted by permission.

Table 5.2 Effects of Terracing -- Indonesia

River	Before Terracing		After Terracing	
	Erosion Rate (cm)	Run-off Coeff (%)	Erosion Rate (cm)	Run-off Coeff (%)
Tapan	2.10	80	0.20	40
Dumpul	2.00	75	0.10	30
Wader	1.40	75	0.20	30

Source: Sunarno, Ir. and Ir. Sutadji. "Reservoir Sedimentation: Technical Environmental Effects." Paper presented to the 14th Congress of Large Dams, Rio de Janeiro, 1982. Proceedings pp. 489-508. Reprinted by permission

Table 5.3 Effects of Reforestation -- Indonesia

River	Before Reafforestation		After Reafforestation	
	Erosion Rate (cm)	Run-off Coeff (%)	Erosion Rate (cm)	Run-off Coeff (%)
Tapan	2.10	80	0.06	20
Pidekso	1.40	75	0.04	25

Source: Sunarno, Ir. and Ir. Sutadji. "Reservoir Sedimentation: Technical Environmental Effects." Paper presented to the 14th Congress of Large Dams, Rio de Janeiro, 1982. Proceedings pp. 489-508. Reprinted by permission.

Although most of the research on sedimentation control has been conducted in developed countries, experiments have also been carried out in a few developing countries. One of them was conducted in the Brantas watershed area in Indonesia where significant sedimentation problems have occurred downstream of a major multipurpose dam. Terracing and reforestation programs were undertaken to determine their effect on sedimentation rates. Sedimentation was found to depend mainly on upstream erosion. As shown in Table 5.2, upstream terracing has decreased the erosion rate from 2.10 cm to 0.20 and the runoff coefficient has decreased by half. Similar results were found in the other tributaries. Even more dramatic decline in runoff were observed in reforestation experiments (Table 5.3).

Impact of Sedimentation on Major Dams

Major Dams of the World

Single-purpose of multi-purpose dams alter the natural environment, particularly the hydrological characteristics of river basins. Among other things, they act as traps, causing backups of sediments carried downstream to the dams. In light of the very high volume of sediments carried by many rivers, a great deal of concern has been voiced regarding the useful life of dams, which are very costly investments (Mermel, 1981).

The Mermel report lists the world's highest dams, the world's largest capacity reservoirs, the world's largest hydroelectric plants and other major dams. The study shows that many dams are located in river basins that show extremely high rates of sediment load. Countries where increased dam construction activities are noted included Brazil, India, Turkey, Argentina and People's Republic of China. The study also shows that dam construction increased substantially during the 1960s and 1970s but seems to have declined during the 1980s and that the water capacity of dams has increased over the years -- that is, countries are constructing much larger dams.

This information has specific implications for sedimentation management schemes. *First*, the decline in the rate of increase in dam construction suggests that the number of readily available dam sites is also declining. *Second*, since dam capacity is increasing, mismanagement of dams will have larger social effects. Since there is a correlation between dam capacity and dam size of catchment area, the data also suggest that dam management will affect larger numbers of people and be subject to a larger number of climatic, geographic and socioeconomic factors. In other words, management of sediments has become more complex and may eventually become an intractable problem. *Third*, assumptions regarding sedimentation rates are extremely conservative.

Water Reservoirs or Sediment Traps

Available data clearly show that the economic life of reservoirs has generally been overestimated (Table 5.4). Actual rates of sedimentation are many times higher than the rates predicted during construction. In India, for example, the assumed rate of sedimentation is 36 times less than the actual rate (Nizamsagar Reservoir in Table 5.5).

As shown in several tables attached, sedimentation is a threat to the economic life of many reservoirs. Given that many dam sites are irreplaceable assets, the costs that sedimentation will impose on future generations (for example, for developing new sources of hydroelectric energy, irrigation and drinking water supplies, controlling floods) will be extremely high. The maintenance of existing dam and reservoir capacity is therefore vital to India's future economic development (Table 5.5).

Table 5.4 Estimated Annual Silt Load at Major Dams

River (location)	Sediment Load (000 m^3)[3]	Dam1	Capacity[1] (000,000m3)	Sediment Load % of Capacity[2]	Source
Nile (Egypt)	128,000	Aswan	164,000	0.08	Salash
Indus (Pakistan)	244,000	Tarbela	13,690	1.78	Khan
Jehlum (Pakistan)	55,327	Mangla	6,358	0.87	Ilahi
Kohat-Toi (Pakistan)	1,423	Tanda	97	1.47	Awan
Kabul (Pakistan)	23,582	Warsak	27	0.29	Holemen
Mahanadi (India)	61,495	Hirakud	8,105	0.76	Holeman
Sutlej (India)	-	Bhakra	9,621	0.35[4]	Gupta
Euphrates (Syria)	4,308	Tabga	14,000	0.03	Holeman
TOTAL	545,624		215,898	0.25	

[1] Mermel, "Major Dams of the World" *Water Power and Dam Construction* May 1981.
[2] Would be the rate of capacity loss if reservoir trap efficiency is 100%.
[3] Assumed conversion factor of 1 tonne equals 1 meter^3, based on AWAN.
[4] Based on observed rate of sedimentation.

Table 5.5 Sedimentation Rates in Selected Reservoirs in India

Reservoir	Year of Enclose	Catchment area (km²)	Annual Average Rate of Silting (ha.m)	Annual Rate of Silting (ha.m/100 km²) Assumed (a)	Observed (b)	Ratio of Observed to Assumed [(b)/(a)]
Bhakra	1959	56876	3421	4.29	6.00	1.40
Panchat	1956	9816	974	2.47	9.92	4.02
Maithon	1956	5206	678	1.62	13.02	8.04
Maynmsagar	1955	1792	360	3.61	20.09	5.57
Niznmsagar	1931	18524	1218	0.29	6.57	22.67
Matatilla	1958	20604	734	1.43	4.43	3.10
Lower Bhawani	1958	4056	110	N.A.	2.70	N.A.
Shivajisagar	1961	777	119	3.42	15.24	4.46
Tungabhadra	1953	27803	1700	4.29	6.11	1.42
Gandhisagar	1960	21873	2200	3.61	10.05	2.78
Hirakund	1956	82652	3178	2.52	3.84	1.52
Machkud	1956	1956	45	3.90	2.33	0.60
Ramgaga	1974	2997	518	4.29	17.30	4.03
Kagsabati	1965	3789	255	3.27	6.73	2.06
Ghod	1966	3629	563	3.61	15.51	4.30
Dhantiwada	1956	2862	181	3.61	6.32	1.75
Ukai	1971	62225	3087	1.47	4.97	3.38
Tava	1974	5983	485	3.61	8.10	2.24
Beas Unit II	1974	12274	1854	4.29	15.10	3.52
Narmada	--	87516	4959	1.55	5.62	3.63
Mahi Stage II	--	25330	2277	1.29	8.99	6.97

- Compiled from capacity survey reports and annual sediment data reports received from project authorities, as reported by Tejwani (1984).
- Sedimentation data are based on capacity surveys in the first 10 reservoirs and inflow/outflow in the remaining 11 reservoirs.
- The studies revealed that the annual rate of siltation from a unit catchment has been 40 to 2,166% more than was assumed at the time of project design (it has been lower in the case of one reservoir only). For the 21 reservoirs, average actual sediment inflow has been about 200% more than average design inflow.

Colombani (1977) carried out several sedimentation studies in Tunisia and found that nearly all the sediment in many basins was deposited behind dams in the basin. These data are consistent with those of CFGB (1980), which reported on sedimentation rates on other reservoirs in Tunisia, Cameroon and Madagascar (Table 5.6).

Table 5.6 Sedimentation Rates

Basin	Transport of Solids Surface (km^2)	Average ($t/km^2/yr$)	Exceptional Year (t/km^2)
Medjerda (Tunisia)			
-at Ghardi Maou	1480	2,000	5600
-at Bou Salem	16230	737	2960
-at Medjez el Bab	21200	850	2940
Zeroud (Tunisia) Sidi Saad	8950	2300	5000
Tsanaga (Cameroon) Bogo	1535	200	--
Sunagao (Cameroon) Nacgtugal	77000	30	--
Mango Ky (Madagascar) Banian	54000	200	--

Source: Comite Francais des Grands Barrages. "Problemes de Sedimentation Dans Les Retenues". Paper presented at the 12th Congress of Large Dams, Mexico, 1976, pp.1177-1208. Reprinted by permission.

Table 5.7 Comparisons of Capacity Lost in Three Major Reservoirs

	D'Iril Emda (Algeria)	K. El Girba (Sudan)	Chabanon Farah (Iran)
Watershed Area (km^2)	657	100,000	55,000
Initial Capacity ($10^6 m^3$)	155 (1953)	1,300 (1964)	1,750 (1961)
Capacity in (year) ($10^6 m^3$)	123 (1975)	840 (1973)	1,400 (1970)
Average Year Supply of Water ($10^6 m^3$)	180	12,000	4,400
Average Year Supply of Solids ($10^6 m^3$)	2.2	65	40

Source: Comite Francais des Grands Barrages. "Problemes de Sedimentation Dans Les Retenues". Paper presented at the 12th Congress of Large Dams, Mexico, 1976, pp.1177-1208. Reprinted by permission.

Demmak (1980) presents data on how the capacity of four major reservoirs in Algeria has been reduced because of high sedimentation rates, highlighting the loss in capacity. The average annual capacity lost at Ksob and Ghrib reservoirs was 4% and 9%, respectively. Demmak concluded that Algeria is losing its potential water reserves at a rate of 20 Hm^3.

Table 5.8 Sedimentation in the Karangates Reservoir Java, Indonesia

YEAR	SUSPENDED LOAD (Hm³)		REMARKS
	GADANG	CLUMPRIT	
1974	0.310	0.375	1. Gadang Station Cover catchment area for 810 km² (total system of 1390 km²)
1975	0.325	0.390	
1976	0.380	0.420	
1977	0.425	0.385	2. Clumprit Station cover catchment area for 494 km² (total system of 660 km²)
1978	0.445	0.435	
1979	0.455	0.425	
1980	0.460	0.461	

Source: Sunarno, Ir. and Ir. Sutadji. "Reservoir Sedimentation: Technical Environmental Effects." Paper presented to the 14th Congress of Large Dams, Rio de Janeiro, 1982. Proceedings pp. 489-508. Reprinted by permission.

CFGB (1976) compared the capacity lost in three major reservoirs in developing countries (Table 5.7). In the case of the reservoir in Iran, 35% of total capacity was lost in less than nine years, equivalent to a sedimentation rate of 4% per year (Table 5.7).

Sunarno and Sutadji (1982) reported that sedimentation of the Karangates Reservoir, on the Brantas River in Indonesia has been increasing since 1974 (Table 5.8). Data from Gadang Station showed that suspended load increased almost 50% in only seven years. This is equivalent to an average growth rate of over 7% per year.

Wu (1984) states that 86,880 reservoirs of different sizes have been constructed in China. However, sedimentation rates are so high that serious damages have occurred. In the Shanxi Province of China, there is an increase of 260 Hm³ of reservoir storage every year, while about 80 Hm³ of storage capacity has been lost because of sedimentation.

Abrasion Effects[2]

Sedimentation may have harmful effects on hydroelectric equipment. Sediments vary in size, shape and hardness and their flow through hydroelectric equipment can result in abrasion of turbines and other steel parts. At the La Florida Hydro plant in Chile, which operated under a head of 95 meters, the turbines were worn out after only 2,000 hours of operation, due to a high concentration of sand in the water. In a hydropower plant constructed in Norway, runner wheels needed overhaul after every 800 hours of operation because of higher concentration of sediments. Annual maintenance costs were U$1.25 million.

Wu (1984) also reports abrasion of turbines caused by passage of coarse particles. "For example, within the 'light' years of operation of the Yanguoxia Hydroelectric Power Plant, the turbines have been repaired five times and 80% of the work was devoted to the treatment of abrasion. Another example is that more parts of turbines have to be changed once every three months in Yilihe Hydro Project in South China." (Table 5.9)

Table 5.9 Sedimentation in Some Reservoirs in China

Reservoir	River	Water shed (km^2)	Capacity 10^8m^3	Sediment 10^8m^3	Sediment %	Annual Sediment 10^8m^3	Annual % of Total
Sanmenxia	Yellow	688,421	97	33.91	44	4.52	5.87
Hongshan	Laohahe	24,486	25.6	4.4	17.2	0.293	1.15
Guanting	Yongdinghe	47,600	22.7	5.528	24.4	0.233	1.01
Qingtongxia	Yellow	285,000	6.27	5.27	86.9	1.055	17.4
Liujiexia	Ditto	172,000	57.2	5.22	9.11	0.652	1.14
Danjiangkou	Hanshui	95,217	160	6.25	3.91	0.417	0.26
Gangnan	Hutuohe	15,900	15.58	1.85	11.9	0.109	0.7
Huanghizhuang	Ditto	7,500	12.2	0.53	4.35	0.106	0.869
Yangguoxia	Yellow	182,800	2.2	1.5	68.2	0.375	17.0
Xinqiao	Hongliuhe	1,327	2.0	1.56	75.0	0.11	5.5
Cetian	Sangganhe	16,900	2.0	1.29	64.5	0.129	6.45
Zhangjiawan	Qingshuihe	8,000	1.19	1.005	84.5	0.201	16.9
Sanshenggong	Yellow	314,000	0.8	0.439	54.9	0.04	5.0
Jiucheng	Luhe	378	0.578	0.5871	100	0.04	7.26
Zhenziniang	Hunhe	1,840	0.36	0.287	80	0.0191	6.7
Zhangjiazhuang	Xiache	460	0.341	0.165	48.5	0.0127	3.73
Gutengshan	Nanyanghe	291	0.228	0.143	62.8	0.0095	4.16
Linghe	Linghe	217.6	0.185	0.1105	60.0	0.0079	4.27
Tuoxi	Zishui		27.30	0.781	2.83	0.0651	0.235
Zhanggang	Lianshui		2.51	0.0424	1.69	0.01445	0.56

Source: Wu, Deyi. "Sedimentation problems in water conservancy in China". Paper presented to the Environment and Policy Institute, East-West Centre workshop on the Management of River and Reservoir Sedimentation in Asian Countries, May 14-19, 1984. Honolulu, Hawaii. Reprinted by permission.

Other Downstream Effects: Floods

Floods are caused by climatic factors (for example, rainfall), by variations in past climatological phenomena (for example, tidal conditions) and by unpredictable events (for example, dam failure). Thus, erosion *per se* does not cause floods. However, changes in land use patterns, particularly practices that deplete the land, are important intensifying factors. Some of the effects are increase in water runoff, decrease in water infiltration, changes in the soil's storage and transmissibility capacity and changes in factors affecting the flowing energy of water. Land use upstream has been correlated with increases in the frequency of floods and with increases in flood damages.

The degree of intensification depends on interactions among climate, topography, geology, soil characteristics, vegetation cover and human influences. As Hamilton (1985) points out, these interactions are complex and are often perceived to be related to the removal of forests. It has been shown however, that there is little cause-effect relationship between forest cutting in the headwaters and floods in the lower basins. The dynamics of watersheds are such that the localised effects of forest clearing (faster runoff of excess water, higher peak floods and earlier peaks) are quickly reduced to insignificance because of other processes. These other processes include the nature and intensity of precipitation, the direction of storms and size and morphology of the basin.

If poor land use in a basin leads to increased soil erosion and sedimentation in flood control reservoirs or in the river channel, flood severity may be increased. Such increased sediment deposition can reduce the ability of dams to impound floodwater or can raise the level of the river bed, thereby intensifying flooding. Flood damage may also increase over time because of the greater monetary value of structures or crops damaged and because larger numbers of people reside in the flood plain.

Although empirical evidence presented below shows that the absolute value of flood damages is not as large as that resulting from other natural calamities, floods have attracted much attention because of their frequency, suddenness and their concentration within geographical areas.

Information on floods in foreign countries has been collected by the U.S. Office of Foreign Disaster Assistance (OFDA) from 1911. Since then, according to OFDA data, 1.2 million people have died in floods, nearly 550 million people have been affected or injured, 30 million people have been made homeless and flood damages have reached almost $23 billion (in current dollars). Framji and Garg (1976) show that the rivers that have had maximum historic flood discharges of more than 50,000 m^3/second, in order of flood magnitude, are the Amazon, Lena, Yenisey,

Brahmaputra, Godovari, Yangtze-Kiang, Mekong, Narmada, Mississippi, Ganga, Irrawady, Ohio, Volga and Amur.

In *Burma*, low lands are prone to floods. Nearly 2 million hectares are severely flooded each year and 3.25 million hectares are moderately inundated. On average, 750,000 hectares under rice cultivation are lost each year due to floods. Large numbers of people are affected as well. In 1965, the Sittang River overflowed, displacing half a million people, and in 1974, all the areas of Central Burma were flooded, affecting nearly 1.5 million people. In addition, these floods destroy property and deplete valuable soil resources. In hilly areas, where shifting cultivation is a dominant type of farming, torrential rains have washed away much of the shallow topsoil.

In *Pakistan*, according to the Environmental Report (1981),

> "it is noteworthy that severe flooding of the Indus plains have been more frequent during the past twenty-five years than over the previous sixty-five years. Some have attributed this change to long-term meteorological conditions. Most observers however, agree that the probable factors have been the gradual denuding of hilly catchments and the introduction of artificial barriers. Deforestation and overgrazing cause rainwater to run-off into the rivers instead of soaking into the ground. The destruction of vegetation also erodes the soil, thereby depositing silt into the riverbeds, raising their levels and permitting them to flow over low-lying flood plains during time of maximum discharge."

Other sources (see, AID 1983) state that nearly 28 million people were affected by floods in Pakistan during the 1960s, 25 million hectares of land were inundated and 4 million structures were damaged or destroyed. Between 1973 and 1978, five serious floods victimized 12.7 million people (1,516 were killed) and the damage was estimated at US$15 million.

In *Bangladesh*, according to OFDA (1983), 151 thanas out of 480 in 13 districts covering an approximate area of 48,000 km^2 are prone to flooding. Thirty million people inhabit that particular area. Deforestation has been suggested as one important cause. A flood that occurred in 1974 killed 28,700 people and caused damages of US$580 million (US$325 million in agriculture, US$150 million in housing, US$76 million in roads and embankments and US$20 million in damage to schools).

In *India* (Tejwani, 1979 and Framji and Garg, 1976), the maximum area affected one year by floods during 1953-1978 was 18 million hectares (in 1976). The average increased from 6 million hectares for the period 1953-1970 to 8.2 million hectares for the period 1953-1978 (42% of cropped area).

The total estimated value of the damages averaged 119,947.9 Lakh rupees (at 1977 prices). The average increased even though investments of Rs 34,766 million in flood protection (covering an area of approximately 8.0 mill hectares) were carried out between 1969 and 1974.

In *Indonesia*, a country of approximately 1 million km^2, there are 1,278 rivers. It has been estimated that the area liable to suffer every year from floods amounts to 750,000 hectares. Statistics for 1976 show that nearly 6,000 hectares of rice fields in West Java and 2,927 hectares in Central Java were damaged by floods. The 1976 floods affected 11,000 houses. The total value of damage was estimated at Rp 175 million.

In *Malaysia*, there have been three catastrophic floods in 1926, 1966-67 and 1971. Severe floods occurred in 1931, 1947, 1954, 1965 and 1971-72. The 1966-67 flood was a 40-year flood which caused damages amounting to $44.1 million, of which $24 million was damages to herds and crops. On average, 10-year floods cause damages valued at $18 million, 1-year floods of $3.0 million.

In *Peru*, there has been extensive forest-clearing activity in the Costa region, which is characterised by steep river valleys. Over grazing by domestic animals has left the soil bare in many areas that are now subject to severe erosion. Flooding in the region has been exacerbated by removal of trees and other plant cover.

In the *Philippines*, average flood damages gradually increased between 1950 and 1970 to P$132 million (in 1971 prices). The minimum was 97 million in 1955, while the maximum is P$173 million in 1960. Flood damage data for seven selected river basins are displayed in Table 5.10.

Table 5.10 Average Flood Damage in the Philippines

River Basin	Area of Basin (km^2)	Area of Flood Plain (km^2)	Population Directly Affected (1960)	Av Annual Flood Damages (million Pesos)
Pandanga	10,540	25,000	1,200,000	16.7
Aguo	7,460	1,720	750,000	6.0
Cagayan	28,110	570	632,000	2.7
Cotadato	20,030	1,730	370,000	4.0
Agusan	11,500	1,870	250,000	3.5
Bicol	3,120	580	670,000	3.0
Ilog Hilabangan	2,100	100	53,000	2.4
Total	82,860	9,070	3,925,000	38.3

Source: Framji, K.K. and B.C. Garg. "Flood Control in the World: A Global View". *International Commission on Irrigation and Drainage.* Volumes I and II. New Delhi, India, 1976. Reprinted by permission.

In *Venezuela*, flood damages in 1969 amounted to Bs 9 million. Most of the damage occurred in the Western and Central Llanos. A flood that occurred in 1970 in the Barcelona area and Zulia State caused damages that amounted to Bs 12.4 million and Bs 25.8 million, respectively.

In *Tanzania*, frequent floods occur in the Rufiji River Valley. OFDA (1981) reports partial crop losses from floods in about one-third of the years since the 1930s. Before then, heavy floods occurred only once every 12 to 15 years. The increase in flood frequency has been attributed to land clearance (for crop cultivation) and destruction of plant cover in the upstream area of the basin.

Estimating Erosion Rates and Sedimentation in LDCs

Several attempts have been made in recent years to estimate average erosion rates and total amount of sediment load. These attempts have focused either on the world at large or on a sample of developed or developing countries. Here, we try to quantify both parameters for developing countries only.

It should be clear that such a global and quantitative assessment will always suffer from important limitations. Quantification is difficult because data sources vary a great deal. Thus, comparisons of our results with those of others may only show the extent to which our estimates are within the ranges suggested by other studies. Despite this limitation, such an aggregate analysis has two main purposes, (1) to create awareness among policy makers who are not aware of the magnitude of these problems, and (2) to relate the total amount of sedimentation to potentially negative downstream effects.

Quantitative approaches can be classified into three groups, (a) "upstream methods," which quantify erosion and sedimentation based on farmers' field data, the Universal Soil Loss Equation being the most popular framework, (b) "midstream methods," which quantify either erosion or sedimentation based on data on sediment loads in rivers and other water-courses, and (c) "downstream methods," which quantify erosion and sedimentation based on volume of sediment deposition in water reservoirs or other infrastructures.

The quality and reliability of the estimates and projections depend upon assumptions about the sediment delivery ratio (or that fraction of the total detached soil that enters a watercourse) and the amount of sediment in a given basin (the bed load) and conceptualisation of the sediment deposition process in major water reservoirs.

The assessment presented here makes use of a midstream method. We decided against the upstream method because most of the variables that enter into the analysis (rainfall, type of soil, topography, vegetation cover) and too environmentally specific and empirical estimates will vary greatly with changes in environmental conditions even within the same catchment area. Also, the downstream method will underestimate erosion rates due to such factors as bed load deposition upstream.

Most of the limitations of earlier studies were due to inadequate data bases, to possible double-counting and to the weighting factors used in aggregating at the regional or worldwide levels. Although our data base is much richer (that is, the sample size is much larger), it still suffers from important limitations. Our data comes from many different sources and was collected using different sampling methods and cover different time periods. While limitations like these are hardly unique to this study, the reader should be aware of them.

The presentation that follows is divided into three parts, (1) past attempts at quantification, (2) the procedures and empirical results of this study, and (3) certain qualifications of the results.

Past Attempts at Quantification

There are numerous studies that present soil erosion estimates for given countries. Some of these estimates are quantified, while others are presented in qualitative form (for example, "severe erosion," "moderate erosion"). It is usually difficult to make use of such estimates because assumptions with regard to key parameters and their corresponding weights, computational procedures and units of accounts are not explicit.

The FAO has estimated that "total historic soil losses" are 2 billion hectares, compared to 1.5 billion hectares now under cultivation. Also, the agency has estimated that 5 to 7 million hectares are irreversibly lost to agriculture each year.[3] This estimate includes land that is degraded by water and wind erosion, salinization and desertification and other factors.

One important study was carried out by two authors whose work was published by the Worldwatch Institute (Brown and Wolf, 1984). Their procedure included the following steps, (a) selection of estimates of excessive erosion in four major agricultural countries (the United States, the Soviet Union, India and China), (b) multiplication of country area by respective erosion rates, (c) division by total area to get a "weighted" average rate of soil erosion, and (d) multiplication by crop area worldwide to arrive at a world-wide estimate.

Their estimates refer to total erosion from cropland and to "excessive" erosion[4]. The estimates are presented in Table 5.11.

Table 5.11 Estimated Excessive Erosion of Topsoil from World Cropland

Country	Total Cropland (million acres)	Excessive Soil Loss (million tonnes)
United States	421	1,700
Soviet Union	620	2,500
India	346	4,700
China	245	4,300
Rest of the World	1,506	12,200
TOTAL	3,138	25,400

Source: Brown, L and E.C. Wolf. *Soil Erosion: Quiet Crisis in the World Economy*. Worldwatch Paper 60. Washington: Worldwatch Institute, 1984. Reprinted by permission.

To obtain a rough idea of excessive soil erosion for the world as a whole, the Worldwatch study assumed that "soil erosion rates for the rest of the world are similar to those of the 'big four'...then the world is now losing an estimated 25.4 billion tonnes of soils from croplands in excess of new soil formation." We have used this assumption to estimate total erosion rates for developing countries. According to our calculations, the Worldwatch estimates for developed countries come to approximately 10.0 tonnes/ha/year, or a total loss of topsoil of nearly 7.4 billion tonnes per year. If one uses their estimates for LDCs only, the erosion rate climbs to 37 tonnes per hectare, or 77 billion tonnes per year.

The Worldwatch Institute's estimates are much lower than those found in other studies. There are several reasons for this. First, the results may be biased since it is possible that erosion from non-croplands might at times be much greater than the erosion from croplands. Second, the Worldwatch results are based on data for only four countries, that is, an aggregation problem.

For the United States the study assumed an erosion rate of 19.3 tonnes per hectare and for India a rate of 37 tonnes per hectare. If one assumes that the erosion rate in India will prevail in most LDCs, the total amount of soil lost will be nearly 300 billion tonnes. Certain aspects of this study should be noted. First, this study used the upstream method, without regard to a sediment delivery ratio. Second, the study distinguishes between total land and cropland and between total erosion and excessive erosion. Third, no weighting factor was used.

Another major study is that by Holeman (1968). Holeman used a midstream method and surveyed data on sedimentation rates in several of the world's largest river basins. Data on suspended sediment loads was organised using an appropriate sediment delivery ratio to estimate the equivalent loss of topsoil in tonnes/ha/year. Holeman's estimated erosion rate was 71.9 tonnes/ha/year. Holeman's data allows for some interregional comparisons. One of the limitations of this study is its use of the number of river basins as a weighting factor. This does not take into consideration the size of the drainage area of each basin.

Jansen and Painter (1974) sought to quantify the role of climatic factors in sediment production using data from surveys carried out in 59 catchment areas in Asia, Africa and Latin America. These catchments were grouped by climatic conditions into 4 categories and statistical analysis was done where a production function approach was used.

The factors taken into account were, (1) catchment area, (2) altitude (in meters), (3) slope of main river channel of the basin (in meters per kilometre), (4) rainfall (in millimetres), (5) temperature (in centigrade), (6) geology (as a dummy variable), and (7) vegetation (also as a dummy variable). Their estimate of total sediment production was roughly 330 billion tonnes, or an average erosion rate of 43.2 tonnes/ha/year.

Methodology and Procedures

This book estimates the total amount of sediment produced and the average erosion rate in developing countries. The results assume a relationship between suspended load data and erosion rates. The study also assumes that the sample of basins was representative and therefore gives adequate results.[5]

Sources of data included Holeman (1968), Sundborg (1983), Brown and Wolf (1984), Prospero *et al.* (1981), Parrington *et al.* (1983), Jansen and Painter (1974), Schumm (1963), UNECAFE (1953), Crosson (1983), FAO/COAG (1981), Ritler (1977), Perkins and Culberston (1968), Chunkao (no date), G.P. Gupta (1975), Yugian (1981), El-Swaify *et al.* (1983), OAS (1974) and Burz (1977).

To implement the quantification procedure, the data either had to be expressed in annual suspended sediment load or allow its calculation. In addition, all the data were converted into metric units. A sediment delivery ratio of 0.05 was used for most catchments, since the corresponding sizes of the drainage areas were larger than 1,000 square meters.

Table 5.12 Comparisons of Soil Loss Estimates for Developing Countries[1]

Source	Sediment Yield (t/km²)[2]	Suspended Load (000 tonnes)[3]	Average Erosion (t/ha)[4]	Total Soil Loss (000 tonnes)[5]
Fournier[6]	570	40,060,740	114.0	869,398,200
Kuene[6]	320	22,490,240	64.0	488,083,200
Gilluly[6]	313	21,998,266	62.6	477,406,380
Pechinov[6]	238	16,727,116	47.6	363,011,880
Schumm[6]	201	14,126,682	40.2	306,577,260
Holeman[6]	182	12,791,324	36.4	277,597,320
Lopatin[6]	124	8,714,968	24.8	189,132,240
Holeman[7]	360	27,454,680	71.9	548,519,050
Jansen & Painter	216	15,180,912	43.2	329,456,616
Worldwatch[8]	49	3,408,677	9.7	73,975,110
Worldwarch[9]	185	13,002,170	37.0	282,173,100
This study	**290**	**20,381,780**	**53.3**	**406,282,780**

[1] Based on a total land area of 76,263,000 km² (FAO, Production Yearbook 1980).
[2] Converted into metric to system: ton/mile²x .35 = tonnes/km².
[3] Sediment yield in tonnes/km² x 76,263,000.
[4] Sediment Delivery Ratio (SDR) of 0.05.
[5] Average erosion times 76,263,000.
[6] Based on Holeman, Table 7, page 746.
[7] Based on Holeman on a regional basis.
[8] Worldwatch, global estimate applied to the area of developing countries.
[9] Worldwatch, estimate for India applied to developing countries.

The estimates are in gross rather than net terms in the sense that movement or deposition of soil within catchments was ignored and that no attempt was made to distinguish topsoil erosion from geologic erosion. Erosion in each catchment was the sum of erosion for each basin, tributaries were eliminated to avoid double counting. Ultimately, data for approximately 60 river basins was utilized. The average erosion rate for a given region was a weighted average, with weights determined by size of the drainage area of each basin.

The most striking result of our computations was the marked difference across regions. While the average erosion rate for developing countries was over 53 tonnes/ha/year, the estimates for Asian, African and Latin American countries was approximately 137, 6 and 12 tonnes/ha/year, respectively. A summary of the results from several studies, including our own, is presented in Table 5.12.

Qualifications

Part of the difference between our estimates and Holeman's may be that his study included fewer river basins and by the fact that rivers are not evenly distributed among regions. Data from the government of Argentina (1970) suggest that our estimate may underestimate the erosion rate for Latin America. Based on these data annual erosion rates range from 1 tonne/ha to 1,000 tonnes/ha. The very low rate of erosion in Africa may be explained by the fact that only water erosion was studied. Prospero *et al.* suggests that between 100 and 400 million tonnes of dust are transported annually by wind from North Africa to South America (Prospero *et al*, 1981, pp. 570-572).

Although the data clearly show that large volumes of soil are being moved from one place to another, it is not possible to correlate erosion with productivity or with the potential costs and benefits of conservation programs. Whatever the losses in productivity may be however, it is clear that the downstream effects of erosion are extremely large in LDCs. A summary of soil erosion losses is illustrated in Table 5.13.

Table 5.13 Soil Erosion in Developing Countries

Region	Catchment (000 km^2)	Land Area (000 km^2)	Area as %	Estimated Erosion (000 tonnes)	Estimated Erosion for Region (000 tonnes)	Estimated Gross Erosion (per Ha)
Asia	7328	26938	28	100,823,840	364,028,420	137.6
Sth America	9995	20201	49	12,383,676	25,049,240	12.4
Africa	10082	29664	34	5,848,460	17,205,120	5.8
TOTAL	27405	76263	36	119,055,976	406,282,780	53.3

Source: FAO Production Yearbook, 1980. Printed by permission.

Notes

1. For a more extensive discussion of this measure see Walling (1983) p 209-237.

2. Derived from Sharma (1973).

3. FAO, Committee on Agriculture. "Soil and Water Conservation," CAOG/81/8, November (1980).

4. That is, topsoil removal at rates greater than that resulting from weathering of the subsoil.

5. The basic data are representative of the basins sampled. Although the aggregation procedure suffered from limitations -- one being that the samples were not representative in the statistical sense -- it is superior to procedures used in other studies.

6

The Economic Nature of Soils

Soil as a Natural Resource

Classification of Natural Resources

Resources can be classified into two broad categories, nonrenewable, or "stock," resources and renewable, or "flow," resources. Nonrenewable resources are those whose total available physical quantity does not increase significantly over a given period of time (for example, coal, oil). Strictly speaking, these resources can grow, but the rate of growth over time is too low to be economically significant. Renewable resources on the other hand can be viewed as reproducible in some manner and can be further grouped into two subclasses. If human actions can decrease future flow, renewable resources are classified as resources with a "critical zone," those whose abundance is impervious to human activity have no such zone. It is therefore important to define this zone.

The "critical zone" is the range below which a decrease in flow, given existing technologies, cannot be reversed economically. Irreversibility may appear much sooner than when the rate of flow reaches zero, as is the case with fisheries and wildlife. This is due to the way in which reproduction of these resources is affected by complex ecological relationships.

Soil is a complex natural resource because it includes both stock and flow resources. Most of the soil's flow resources have a critical zone of exploitation, below which irreversible damage may occur. These resources are affected (that is, consumed) by agricultural production systems, which make demands through plant growth but which may also renew some resources. Most scientists find it convenient to measure the quality of soils in terms of land productivity or crop production. As mentioned earlier, loss in productivity occurs because of reduction in root-zone depth, losses in plant nutrients, degradation of soil structure and the like.

Because of the composite nature of soil resources, an increase in productivity is not necessarily correlated with soil conservation and the lack of such a correlation presents particular problems. A decrease in soil productivity due, for example, to mismanagement of a flow resource (for example, plant nutrients) may be reversed by economically viable activities (for example, application of nitrogen fertilizer). However, resource use may be such that when the critical zone becomes evident (for example, existence of gullies), a return to the original or even an acceptable level of soil productivity may become economically prohibitive.

Shaxson [1981] distinguishes between soil conservation -- building up nutrient capacity -- and soil reclamation -- bringing soil back to its original state by improving such qualities as texture. He argues that this distinction is important in designing soil conservation programs. A soil reclamation project, he argues, is more expensive than soil conservation and requires engineering or mechanical work. Soil conservation, on the other hand, involves biological methods.

Distinguishing between renewable and nonrenewable resources is important not only because soils are a composite resource. Three other implications are worth noting. One is that studies that focus on assessing changes in productivity due to erosion have either considered soil as a uniform resource or as composed of several elements (some renewable and some nonrenewable). Five types of studies can be distinguished. First, there are those that focus on yields without identifying the type of resource that is being depleted (or the one whose critical zone has been surpassed) and is thus the main cause of any change in yields. Most of these studies are experimental in nature and are carried out under controlled conditions. Second are those that focus on yield changes as these are affected by depletion of one or more renewable elements. Good examples are studies focussing on the role of plant nutrients and those that assess the effect of "neutralizing" imbalances among the chemical components of the soil (for example, experiments with acid salts or where the pH factor needs correction). Third are studies focussing on a nonrenewable element, such as soil texture or structure. Fourth are studies focussing on combined effects, such as those that correlate changes in yields with changes in topsoil (for example, soil depth). Fifth are studies that focus on any combination of first four.

A second implication to be noted is that attempts to distinguish between types of soil elements have raised a great degree of controversy. One school of thought sees every soil element (or resource) as renewable and believes that the issue is to determine at what cost this renewability could take place. Another school of thought holds that some soil elements are nonrenewable, thus, soil use is equated with the mining of minerals. In a later chapter we will show that the first view is the rationale for some economic approaches and that this distinction has important implications when assessing farmer behaviour.

Finally, from an economic perspective, there are several renewable resources that can be replaced or compensated for. Most often, the resources that can be replaced are traded in the market (for example, nitrogen fertilizer). Because of this, quantification of benefits and costs can be done in monetary terms. Nonrenewable soil elements, on the other hand, are often non-traded. This poses a challenge when it comes to assigning monetary value to the benefits and costs of soil conservation programs.

A Digression

Sediment may also be defined as a natural resource. The deposition of sediments may have positive as well as negative effects.[1] In this section we are only concerned with the negative effects.

Sediments possess several characteristics which are of interest in determining the effectiveness of sediment management programs and policies. First, sediments are very abundant and have different qualities. These differences however, are seldom taken into account in the design and appraisal of development projects. Second, while other resources are "produced" by the same unit that will make use of them (or sell them), sediments are a by-product of one unit in the economy and are "consumed" or "stored" by other units. Third, the number of production units of the typical stock resource is often smaller than the number of consumer units, whereas here the number of production agents is very large compared to the number of units they affect (for example, dams). Consequently, there is a "concentration effect" downstream. This is an important fact to keep in mind, particularly in dealing with such issues as cost recovery, law enforcement and agricultural incentive or compensation schemes.[2] Fourth, the stock often has more value *in situ* (on the soil) than it has when detached and transported to another part of the system. Because of this, it often makes economic sense to carry out preventive types of conservation programs (for example, reafforestation, gully control structures). Fifth, the "extraction" cost is very low and the regulation of extraction rates depends on how the originating units use other resources. Consequently, there are no incentives to curtail sediment production unless individual producers have to pay for its negative external damages downstream. Sixth, this resource is rarely traded in the market, although changes in volume, size, or chemical concentration affect the productivity and market value of other tradeable commodities. And seventh, sediment are a "mobile" resource. This is particularly important in areas or countries which lack upstream management policies.

Because of these characteristics of sediment, their production and distribution through soil erosion are a clear example of the on-site, off-site typology of soil erosion effects mentioned earlier. By formulating a two-by-two matrix, all of the effects can be divided between those that occur on-site and those that occur off-site and between those that are traded (and have a market price) and those that are not traded (and are thereby more difficult to value). This simple division highlights two major problems -- the *location of the effect* and the question of *monetary valuation.*

As seen in Table 6.1, the range of effects can be divided among four quadrants, those that occur on-site and have market prices (Quadrant 1), those that occur off-site and have market prices (Quadrant 2), and those effects that do not have easily observed market prices, either on-site (Quadrant 3) or off-site (Quadrant 4). Whereas a private financial analysis of a soil conservation project would tend to focus on on-site market effects (Quadrant 1), a socioeconomic analysis would look at all four quadrants and attempt to evaluate and include as many effects as possible. The inclusion of on-site non-market effects as well as off-site effects, both market and non-market, is an attempt to internalize most of the traditional externalities of a soil conservation project. This type of analysis also points out how many individuals are involved in each step. Because soil erosion is commonly the result of the actions of many individuals (usually small farmers), the development of soil conservation projects requires explicit consideration of institutional factors and financial incentives.

Since it is rare for soil erosion and its consequences to occur solely within one individual's or organization's property, institutional arrangements and incentives must be considered to assure successful implementation of soil conservation projects. Even if costs are centrally funded, actual implementation may require coordination among various institutions and organizations in the watershed. In the past however, it has frequently been difficult to achieve coordination. Examples of different government units working at cross-purposes (for example, one unit promoting better management of agricultural lands while another unit promotes population in-migration and the creation of new agricultural areas in the same water-shed) are not hard to find.

Decisions on how much to invest in soil conservation depend on how much society values *future* uses of soil resources. If society decides to invest heavily in preserving soil resources, this decision is equivalent to redistributing income to future generations.

Resource Management Principles

Management of natural resources in general and of soil resources in particular, is essential if development is to be sustained in the long term. The analyst of soil conservation projects therefore needs a set of principles that can be translated into guidelines for specific programs.

Operationalization of programs requires an understanding of commonly accepted principles of natural resource conservation and management. If there is no agreement among the parties involved in the program on the principles of conservation and management, any action that affects the natural environment will be carried out in ignorance of its consequences.

Many professionals in the field however, hold the view that there are no conservation or resource management principles general enough to be applicable to every country, or to every region within a country. They believe that every situation is so environmentally specific that it must be treated as a special case. This is not necessarily true. Here, we try to define resource management principles that are applicable to the management of both nonrenewable and renewable resources regardless of region, country, or program.

This should not be taken to mean however, that the principles can abe applied without being modified to take local conditions into account. Terms like "sustainable development," "resource conservation," "rational management," "renewable resource," "regenerative capacity of the environment," and "genetic diversity" are often used without being defined within a broad policy context. This creates confusion among policy makers, as well as problems during program implementation. Confusion results when it is never made clear what these concepts really mean, or because they mean different things to different people. Problems during program implementation are created because these concepts do not have clear operational significance and therefore become meaningless to project planners and supervisors.

An attempt will be made here to outline the basic meaning of "conservation," (or "depletion") recognizing that there are ethical connotations of the term.

The Meaning of Conservation[3]

Conservation, or the countering of depletion, is a dynamic process which reflects changes in the rate at which a resource is used over time. When any given natural resource is exploited, a redistribution of use rates toward greater use in the future leads to conservation (a decrease in present use). By the same token, a redistribution of use rates toward the present leads to depletion (an increase in present use rate). Thus, the study of conservation with respect to existing natural resources consists of comprehensive analysis and comparison of redistribution -- that is, the direction of change of use rates over time. Conservation is a dynamic concept and does not mean nonuse of resources.

Conservation of soil resources needs to be defined in terms of an intertemporal redistribution of use rates, including the use rates of both renewable and nonrenewable soil elements. In other words, soil conservation means both the protection of the quantity of soil and also of the quality of that soil. It is difficult to measure the state of conservation of soil resources quantitatively, since this would require measuring the redistribution of several types of use rates and also assessing complex interactions between stock and flow resources (for example, nutrients, soil texture). These measurements become even more complex when one notes that some plants or cropping systems may "conserve" soils under one system and "deplete" it under another system. Several interrelationships become evident chiefly through their effects on costs, or on the ability of soils to produce revenues from production. Complementarity between resources -- for example, supply (the extraction of minerals) and demand (the use of coal and iron ore to produce steel) -- and resource competitiveness, whether in supply (for example, agriculture and forestry) or demand (for example, fuelwood and oil), must be understood.

To determine the critical zone of soil resources, planners need to monitor soil quality. But measurements of soil quality are often clouded by a commonly held view of the relationship between conservation and agricultural investments. Investment is normally said to be a cause of depletion (for example, an increase in livestock investment depletes pasture lands) while disinvestment (for example, a reduction in stocking rates) is believed to lead toward conservation. This is not true in all cases, however. This book focuses on the types of investment and institutional change that will shift the distribution of use rates toward the future or that will avoid irreversible damages (that is, violate the critical zone of a renewable soil factor). In addition, a redistribution of use rates may also result from change in institutional arrangements (for example, tenancy, property rights, prices, taxes, subsidies), even when the investment rate is held constant.

More research should be carried out to determine the critical zone of soil resources. Stevens (1978) defines the "critical zone" for soil as the point at which nutrients become unavailable for plant growth (in land cropped year after year without being fertilized). Beyond this point, it becomes uneconomical to rejuvenate the land. The critical zone depends on the environmental characteristics of the soil.

In appraising development projects, "conservation" or "depletion" effects can be defined in quantitative terms as the results of all conceivable alternatives "with" and "without" the project.

Conservation Decisions

Since there are many reasons why use rates may tend toward soil depletion, conservation decisions are complex. Individual farm households are the decision-making units at the micro level. One can assume that their decisions depend on income, institutions and a planning horizon (time). Low-income farmers, for example, will be less willing to postpone consumption -- and thus ease the pressure on intensively cultivated land which are quickly eroding -- than will higher-income farmers.

Changes in farmers' behaviour can be caused by changes in biological, technological, economic, social, political, or institutional factors. Farming systems show the effects of unbalanced changes through such actions as shifting cultivation, range burning, lack of fire protection, overgrazing and uncontrolled tree cutting, all of which deplete soil carrying capacity.

The word "institution" is used in this book to mean "a social decision system that provides decision rules for adjusting and accommodating, over time, *conflicting* demands (using the word in its more general sense) from different interest groups in society" (Ciriacy-Wantrup, 1969, p.131). Institutions affect the use of resources as well as the distribution of income derived from that use. There are market and nonmarket institutions. Market institutions include such things as prices, interest rates and taxation. These factors work through the market. Nonmarket institutions include such things as property rights and tenancy.

Optimality in Resources Management

We have said that countries can establish resource management principles that apply to both renewable and nonrenewable resources and that those principles are general enough to provide a policy framework for investment planning and implementation decisions. *Environmental policy is concerned with achieving an "optimal" state of conservation* within this framework. Therefore, this state can only be approximated.

Ways to achieve this optimal state are the subject of this section. The notion of optimality has always been important to economists. Economics textbooks often define economics as being the science concerned with the allocation of scarce resources to satisfy a very large number of needs. Scarcity -- the inability to satisfy all needs with the limited amount of resources available at a given point in time -- is at the heart of economic decision making. Optimal allocation means that existing resources should go as far as possible. Thus, decision making will be guided by the needs to be satisfied.[4] One then encounters the problem of distinguishing between "private" optimization and "social" optimization.[5] The first type of optimization satisfies an individual farmer's (or other party's) objectives, while the second type satisfies society's longer-run objectives. Conflicts often arise when individual decisions do not collectively achieve socially desired goals. In other words, what seems right to the individual farmer may not necessarily seem desirable to society. A comprehensive treatment of resource management policies involves addressing the definition of optimality and also considering policy implementation.

Optimality and Decision Making

In this section we characterize private optimization and social optimization in the context of soil conservation practices. *Private farm units* customarily seek to maximize profits (or net revenues) within a given set of constraints. These constraints include the farm unit's factors of production (for example, capital, labor), the policy environment (for example, market and nonmarket factors), time and uncertainty. Cultural values and attitudes are also part of the decision environment, whether as stimulators or as constraints.

Changes in institutions or economic factors may change the course of action taken by an individual farmer in favour of either conservation or depletion. For example, a price support system that artificially increases the price of a commodity whose production will lead to depletion -- other things being equal -- will provide incentives to farmers to grow more of that crop and consequently, to deplete the soil. Taxes on inputs, on the other hand, will have the opposite effect. Low interest rates often tend to shift use rates towards depletion. If real interest rates are artificially held below equilibrium, as is common, borrowing will increase. Hence, revenues will be distributed toward the present and costs toward the future. Incremental benefits today tend to be considered to be worth much more than net benefits in the future.

Uncertainty is an important factor in regard to soil conservation decisions. There is uncertainty with regard to such things as tax levels, future commodity prices and the weather. Farmers must constantly make decisions about the allocation of available natural resources like soil and water and other resources like capital, farm inputs and labor. Factor endowments and technology change over time effect the way in which private returns are accrued.

Consumer preferences may also change over the long term. Suppose, for example, that a forestry program is initiated to prevent erosion and local residents are provided with coal as a substitute for firewood. Over a long period of time this change will be accompanied by technological changes (the installation of kerosene or coal-using stoves). As a result, local residents may never go back to using wood as a source of energy.

Social optimization refers to whether society allocates resources optimally and thus achieves an "optimum" state of conservation. Achieving such a goal depends on the extent to which society's views differ from the views of individuals. Empirical evidence shows that these views differ with regard to the conservation of soil resources.

The obvious questions that follow are, What should society's resource management objective be? and -- what guiding principle should society use in managing its natural resources? These are difficult question. Ciriacy-Wantrup [1968] offers one of many proposed guiding principles: The *Safe Minimum Standard of Conservation (SMSC)*.

The exploitation of flow resources raises the risk of irreversibility. The more irreversible damages this generation produces, the more we narrow society's future options. The Safe Minimum Standard of Conversion is achieved by avoiding the "critical zone", the point at which it becomes uneconomical to try to reverse depletion.

The use of such a principle as the SMSC will increase society's flexibility and options. On the other hand, of course, certain costs are involved in maintaining such a principle. If soils are heavily eroded, the application of the SMSC may require foregoing the use of land for some time. This is land where the use rate is so high that soil depletion may be unavoidable unless drastic actions (for example, resettlement of people in other locations) are taken. In other situations, it may be sufficient simply to adopt certain conservation practices.

Meaningful application of the SMSC requires detailed analysis and knowledge of soil resources. In practical terms, one may define the SMSC for soil resources as avoiding the formation of gullies, or, more generally, as the maximum possible rate of soil erosion under a given set of economic and institutional considerations. The SMSC can be translated into farming practices (for example, contour ploughing, minimum tillage, mulching) and other actions designed to avoid the critical zone.

Policy Implications

Domestic policies on training, extension services, research, zoning, compensation, regulations and public land management in general may be needed to achieve the SMSC. The effectiveness of these policies depends upon the economic, institutional and social environment of the country.

Because of the nature of soil erosion, efforts to control the problem may not be as important as efforts, for example, to manage fisheries resources. But floods, sediment and cloud dust from one country may affect other countries. In addition, important instruments may indirectly affect the rate at which soil resources are being depleted. These include commodity or trade agreements, management of water resources, technology transfers and educational and institutional changes. Some coordination of soil erosion programs at the regional, national and international levels is therefore required.

After principles and instruments are identified, research is needed to evaluate the effectiveness of alternative policies.

Other Economic Issues

Since most of the issues outlined earlier and in the following chapters have some economic connotations, this section deals mainly with micro/project-related problems. A good way to start is by asking the following question: Is there any characteristic that makes soil conservation programs unique when compared with other programs, such as industrial projects? The economic issues involved in the design and evaluation of soil conservation projects are, with few exceptions, similar to the economic issues involved in any other type of development project. These exceptions, however -- the existence of important economic externalities and associated major valuation questions -- pose problems in carrying out an economic analysis of soil conservation programs. Since many of these externalities involve potential benefits (most of the direct *project costs* are known with much greater precision), the tendency is to undervalue net benefit. As a result, many soil conservation projects appear "unattractive" in conventional economic terms.

Impacts on Equity

Soil conservation projects, by changing the current and future characteristics of erosion, affect on-site and off-site economic activities. Conservation requires changes in farm practices, this, in turn, alters the costs of production and often, the level of production. Because of their long term nature, soil conservation projects redistribute welfare, not only among economic actors of this generation but also across generations.

This redistribution means that economic analysis must be carried out in a way that makes it possible to assess the major tradeoffs between the welfare of different groups and generations. This analysis must deal with four basic questions, (1) What measure should be used for benefits? (2) How should welfare impact be measured? (3) How should welfare measures be aggregated? (both intertemporal aggregation and interpersonal aggregation) and (4) How should welfare impacts be presented when they cannot be given a money valuation?

External Effects

Many of the effects of soil erosion caused by one individual are felt somewhere else in the environment. This fact is especially evident when soils in the form of sediments affect economic activities downstream, such as hydroelectric generation, irrigation agriculture and drinking water. These economic externalities impose costs on others and thus decrease welfare. Placing a value on these effects is an important part of the analysis of soil conservation projects.

Time Effects

Economic analysis of any project involves dealing with time. Benefits and costs are distributed over time and measures of project worth are reduced using indicators of time dimensions. The most common procedure for measuring worth is "discounting," where a common yardstick is developed based on a given discount factor. Thus the present value (PV) of a flow of benefits and costs over time is given by,

$$PV = \frac{B1 - C1}{1 + r} + \frac{B2 - C2}{(1 + r)^2} + \ldots \frac{Bn - Cn}{(1 + r)^n}$$

Where B1 and C1 are defined as benefits and costs in year 1, r is the discount rate and PV is the present value.

When valuation in the absence of markets is attempted, attention is focused on how to arrive at the values of B and C when the analyst does not have revealed values or prices. As the time span lengthens, or as projects have more net benefits in the future, net benefits decrease more substantially because the discount factor increases exponentially.

Some of the impacts of erosion are instantaneous, while others are distributed over time. The basic nature of soil erosion control programs can be understood if we point out that the negative C element in the above equation (that is, the costs of a soil conservation program) pertains to the present -- with a high weighting factor -- while the positive B element (that is, the benefits of a soil conservation program) pertains to the future and has a low weighting factor.

For example, a rate of one millimetre of erosion per year may not greatly affect farm productivity over the next 75 years or so, but after 100 years it may render farming of a particular piece of land impractical forever. The same principle applies to off-site effects. Sediments may collect in a reservoir at a very slow rate and only change the practical value of the reservoir some 50 to 100 years in the future.

Thus, intertemporal consequences are an important characteristic of soil erosion control programs. Current soil conservation practices cannot ignore the welfare of future generations. If farmers decide not to take actions to conserve soils, they in effect impose severe costs on the welfare of future generations.

Whenever a positive discount factor is used in computing net present value, the welfare of the current generation is to that degree favoured over future generations. However, decisions with regard to balancing the welfare of present and future generations involve moral value judgements. But it is only the discount factor (or the denominator of the NPV equation) that matters in this calculation. Also the analyst will necessarily be uncertain about the future values of such outputs as food, land and reservoir capacity.

Several questions should emerge from this discussion: Should the analyst use one or several discount rates? How should alternative discount rates be adjusted for in the analysis? How can society persuade farmers to act in accordance with society's discount rate, which will be smaller than the discount rate which farmers obtain? And how can lending agencies justify investments in soil erosion projects which may have a relatively low rate of return?

Valuation in Absence of Markets

Soil erosion changes the quantity and the quality of both inputs and outputs of different economic activities. When markets exist, changes in the quantity and quality of products are often reflected in price changes through the normal process and supply and demand. As mentioned earlier however, there is no market for such things as polluted (or unpolluted) river water, sedimentation, or residual concentrations of nutrients or chemicals.

Consequently, valuation in the absence of markets is a major concern (Sinden and Worrell, 1979 and Hufschmidt *et al.*, 1983).

Risk and Uncertainty

The future is uncertain, one does not know the probability of many events. This uncertainty extends to estimates of future prices and future demand. Exposure to risk and attitudes toward risk, therefore become important issues at the farm level. Soil conservation projects may not only raise costs and alter the level of expected income but may also increase the variability of that income. New technology or farming practices may mean that the increase in farmers' gains will be much higher than it would be without soil-conserving practices. A basic question, therefore, is how large a premium is needed to get farmers to adopt a particular soil conservation practice.

Multiproduct Nature

Soil conservation programs often have other effects besides their impact on the ecology of soils. Forestry programs, for example, may produce fuelwood, timber, fruit, woodpoles, fodder, water catchment protection, shade and improved flood control.

The nature and structure of demand will influence the extent to which a program is effective in conserving soil. High relative prices for energy, for example, or strong population pressure, may result in an increase in demand for fuelwood. This higher demand will translate into a higher rate of forest depletion, with negative ecological effects on existing soils. But if a forestry project is under consideration, higher demand will raise the value of anticipated benefits.

Micro-Market Linkages

Soil conservation programs potentially affect production factors and output markets. The potential effects are not unusual except for their influence on land markets. Soil erosion affects both the quality of land as a flow resource and total amount of land as a stock resource. The effects are a unique characteristic of soil conservation programs. From an economic standpoint, changes in the rate of erosion will therefore not only change the flow value of land (that is, the present value of the income streams attributable to the land, which is capitalized into the value of the land), but also its stock value for society.

The main issue is therefore society's valuation of land as opposed to farmers' valuation of land. The need to control erosion reflects the fact that the value of land as perceived by the user is different from the value as perceived by society. There are several possible reasons for the difference:

1. Farmers do not realize the long-term economic consequences of erosion.
2.The market misjudges the future demand for food and fibre and thereby underestimates the value of land.
3. The market over-estimates the rate of development in erosion control measures is less than the private cost.
4. The maintenance of land to ensure ample supplies of food and fibre is valued more highly by society than by land users (Crosson and Brubaker, 1982).

A major issue then, is how should society try to preserve its land use options as a hedge against future demands for food and fibre.

Soil conservation programs may influence product prices in a general way. If soil conservation measures are to be effective in controlling erosion, they usually must be implemented in a large scale, if not in the country as a whole.

Large-scale programs may change the output mix, the level of production and the products that enter the market. Economic analysis therefore cannot be carried out on the assumption that prices will remain constant. With regard to markets in general, there is a question as to whether the price reflects the true value of commodities at the farm.

Irreversibility

An important consequence of soil conservation programs is the possibility of irreversible consequences. This possibility complicates the process of valuation, how should irreversibly lost resources be valued? The final value of land will depend on the extent of the remaining stock of land.

Implementation

A crucial but frequently overlooked issue is the institutional arrangements that may be needed for successful implementation. Direct feedback of benefits to those carrying out the project to ensure successful implementation cannot be relied on.

Almost all the costs of a watershed stabilization program will be incurred in the uplands, but most of the benefits will accrue to downstream activities.

Table 6.1 The Location and Valuation of Soil Erosion: Related Effects

		Location of Goods and Service	
		On-Site	Off-Site
Types of Goods and Service	Marketable	I	II
	NonMarketable	III	IV

Examples:

Quadrant 1: Changes in crop yield, change in input use, damage to farm infrastructure.
Quadrant 2: Changes in agricultural production, damage to infrastructure, changes in water quality (and costs associated with these changes).
Quadrant 3: Changes in the appearance of the land, change in soil structure.
Quadrant 4: Aesthetic changes in land, aesthetic changes in water quality or quantity, changes in species or genetic diversity

While economic analysis includes benefits and costs wherever they occur, actual implementation will required special incentives to make private benefits coincide with social gains.

A second major implementation question is the different between watershed areas and the territory supervised by administrative units. While a watershed has a natural size that is largely determined by gravity, geographical-administrative units are rarely identical with watersheds.

Notes

1. In ancient times the Romans built dams for the sole purpose of trapping sediments. These sediments were then used to expand agricultural lands. Sediments are very fertile, and many areas of the world could not be cultivated without them. In Africa, many farmers cultivate the land near the edge of lakes during the dry season to take advantage of the wetness and fertility of these soils.

2. Sediment characteristics present special problems for effective sedimentation management. Investment decisions by units "consuming," "storing," or "eliminating" sediment depend on the outputs of other units which are not under their control. Furthermore, the producing sector is

a "leader" and the receiving sector is a "follower," which often results in larger and more costly investments or in higher marginal or total costs of production (e.g., dredging of ports, abrasion of hydro-electric equipment).

3. The meaning of conservation, and the nature of conservation decisions, are the subject of many books. The interested reader should refer to the bibliography.

4. The concept of resource allocation focuses on users, while the concept of conservation focuses on optimality with respect to time (Wantrup, 1971).

5. One of the main characteristics of soil conservation programs is that the time distribution of benefits and costs differs. The distinction between private and social optimization is important for at least three reasons: (1) because the resources that an individual farmer can devote to soil conservation practices may be much less than what is thought to be socially desirable; (2) because environmental resources in general, and soil resources in particular, have a much greater economic significance for society than for the individual farmer (i.e., society has to plan in perpetuity); and (3) because efficiency maximization at the farm level may differ from what benefits the country as a whole. The reader will find an analysis of these factors in Wantrup (1947) and the comments by Nelson in the same issue.

PART THREE

Introduction

The major objectives of Part Two were to define the soil erosion problem, to create awareness of the estimated magnitude of soil erosion effects in developing countries and to conceptualize soil as an economic resource. This Part provides analytical frameworks and methods to improve the design and appraisal of soil conservation programs.

An analytical framework provides the basis for investment decisions. This framework identifies the most important technical and economic relationship, presents alternative methodological approaches in relation to a theoretical foundation and applies the methods, as shown in a series of illustrations.

Chapter 7 outlines the necessary steps needed to integrate economic analysis with technical analysis of soil erosion effects. Often, the role of the economist begins where that of the natural scientist ends. There is a large gap between the work of the economist/financial analyst and that of natural scientists. Communication and understanding of their respective roles and of the scope and limitations of their analytical frameworks, are essential for successful implementation of appraisal missions.

Chapter 8 presents the technical frameworks that are most often used to characterize such natural phenomena as erosion, sedimentation and floods. Instead of explaining the specific models which hydrologists, soil scientists and agronomists use in each case, the chapter focuses on analyzing the relationships that influence technical decisions. Much effort is devoted to explaining the technical nature of the most important on-farm downstream effects (for example, sedimentation, floods of soil erosion). Reference is also made to other types of downstream effects.

Chapter 9 begins a sequence of chapters on specific economic issues that arise during the appraisal of different aspects of soil conservation projects. This chapter provides the theoretical foundations of economic appraisal methods and outlines alternative ways of dealing with valuation methods -- for example, how to price commodities that are not directly traded in the market place. With regard to the theoretical foundations of appraisal methods, this chapter points out that many analysts lose sight of the fact that computations of any measure of project worth have to conform with the basic principles of welfare economics. For example, economists very often decide to quantify soil erosion effects by estimating the replacement value of nutrients lost due to erosion, which is then used to compute the project's net incremental benefits to the economy. However, many of them do not know why that particular measurement is, in fact, a true notion of welfare. One needs to be aware of several assumptions that are used when making these welfare estimates. These are explained in some detail. Chapter 11 deals with certain aspects of the economic analysis of projects -- for example, shadow pricing. The reader who is not familiar with shadow pricing, with how to deal with inflation, with the meaning of rate of return, or a net present value estimate, should consult textbooks that deal with benefit cost analyses. Finally, the last part of this chapter analyzes the many ways by which economists arrive at a monetary value of nonmarket benefits or costs. This section proposes a typology of valuation methods that is used later on.

Chapters 10 and 11 present several economic approaches to the appraisal of upstream and downstream effects. The approaches differ in the assumptions made with regard to such parameters as cropping pattern, technology adoption, impacts of sedimentation, price variations and several other factors. Each method is presented in detail, including major assumptions, data requirements and possible limitations. These chapters do not present all possible methods. After one has an understanding of the theoretical foundations and practical implications of the methods presented here, one may be ready to expand or modify such methods. Methods deemed impracticable in developing countries have been excluded.

Chapter 12 presents some case studies, or illustrations, of how some of the methods analyzed in Chapters 10 and 11 have been applied. In many instances the presentation is not comprehensive, since the material used was taken from the literature or reports, few of which were drafted in case study form.

7

Integrating Alternative Frameworks

The Need for Integration

Soil conservation projects involve the management of natural systems (for example, soil, water, forests). Descriptions and analysis of these systems are the subject of several disciplines. These systems are complex and attempts to achieve balances among climatic, hydrologic, geologic and other natural factors present policy-makers with difficult decisions.

Intermingled with these natural systems are economic and social systems. While the social systems relate to how people are organized in society in a pursuit of certain objectives, the economic system deals with the allocation of scarce resources in satisfying human needs. Decisions regarding socially desirable investments are frequently made by policy-makers. All investments in one way or another affect the other systems. It is of particular importance to note however, that in the field of soil conservation, investments affect natural systems (or the environment). Consequently, the identification, preparation and appraisal of soil conservation programs involves the integration, in conceptual terms, of the natural system with the economic system. This does not mean that integration with other systems is not needed. On the contrary, programs that do not take the social or political environment into account are bound to fail.

This part of the book deals mainly with the conceptual integration of the natural and economic systems. The next part of the book outlines issues relating to other decision systems. The basic subject here is integration of the approaches used to assess the social desirability of investments in soil conservation.

To establish how the integration between technical frameworks and economic methods should take place, one has to define first the "focal point" of integration. A useful way to address this problem is by asking where analysis by the economist begins and by defining the extent to which technical frameworks and data (parameters) are useful in carrying out the economic analysis.

As stated in the previous chapter, this integration has seldom taken place. Thus, the work of appraisal missions tends to be fragmented. If one is to improve the quality of investment decision analysis, the data generated by the natural scientists has to be connected with the departure point of economic analysis. The departure point of any economic analysis is the notion of productivity. Therefore, the data generated by technical frameworks must be related to the notion of productivity. Otherwise, technical research is immaterial to the economic analysis of a project. Up until now, technical investigations have rarely provided usable data for economic analysis.

Likewise, if the economist does not have a clear understanding of the basic relationships underlying the functioning of a given natural system, it will be very difficult, if not impossible, to identify benefits and costs or to formulate a notion of productivity.

Let us begin with the problem encountered when one is trying to determine losses in soil productivity due to erosion. The work of the economist begins with the establishment of some notion of land/soil productivity based on yield projections which have been estimated by the agronomist. These estimated yields depend upon the volume and characteristics of existing soil. The economist, using prices, output levels, cropping patterns and the like, will then try to simulate behaviour using farm budgets and then aggregate them to estimate overall economic return. It is important that the economist understands the relationships and assumptions behind the yield figures provided by the agronomist. As explained in Part Two, the relationships between erosion and losses in yields are complex. These need to be clearly understood if proper interaction within the appraisal team is to take place.

Floods often occur because of an increase in runoff or sedimentation. When floods affect agricultural land under production, the economist needs to know how the hydrologist defines the main ingredients that go into a computation of a "flood damage factor". This factor is a basic input for economic analysis and its computation requires knowledge about the probability of floods, types of floods, water depths at different points in time and several other characteristics. The agronomist uses these flood characteristics in determining the type of productivity losses that occur as a result of floods. Where property damages occur, inputs from other qualified professionals are also required (for example, real estate assessors to establish losses in property values). Only after these factors have been computed and taken into account can the economist begin the economic analysis.

It should be remembered that some level and frequency of flooding is a natural event that would occur even without human activity. This fact does not decrease the "costs" of flood damage or the "benefits" of preventing that damage. Rather, it demonstrates that even a complete elimination of man-induced erosion and sedimentation would not guarantee harmless floods. Soil conservation measures should however, decrease the frequency and severity of floods.

This chapter outlines in a step-by-step manner what is needed to achieve a higher degree of integration and it will give a specific example of how this integration has taken place.

Seven Steps for Integration

Integration may be achieved at the level of an individual project effect (for example, integrating economic and hydraulic frameworks to estimate the effects of floods), or at the level of the whole project. The approach here is rather broad, since it is followed in the next few chapters by a detailed presentation of alternative frameworks.

The first step toward the integration of frameworks consists of proper identification of soil erosion effects. Experience shows that many soil conservation projects have been appraised only partially, in that not all the potential effects have been considered. This often happens when no consideration is given to downstream effects because a soil erosion control project is conceived as being only an agricultural sector project.

The second step consists of describing the physical characteristics of the phenomena under consideration. This step is required to properly identify the major causes of erosion by defining cause and effect relationships. Thus, several technical relationships will have to be defined with their accompanying physical flow tables. In this step the basic characteristics of erosion, sedimentation, floods, water quality problems and the like, are specified. These, in turn, are linked to the point of departure for the economic model. For example, on-site effects of erosion may need to be related to soil productivity, perhaps through modelling reductions in available water (or nutrients) and its effects on plant growth and yields. The number of areas to be covered will be determined by the type of erosion effects defined in the first step.

The third step consists of defining the basic units of account that will be used by the appraisal team to analyze soil erosion effects with and without a soil conservation project. For example, the appraisal team may decide to use as units of account a yields estimate, a water quality indicator, a volume level of water supply, an estimated level of flood damage (in physical terms) and the like. It is at this stage that the departure from natural science into economic analysis takes place. These physical units will be the basis for assigning values to the benefits and costs associated with the project.

The fourth step consists of defining the type of welfare estimates the economist plans to use, based on the type of physical parameters defined earlier and including other economic or social variables (for example, prices).

The fifth step consists of developing value estimates to quantify welfare gains or losses, comparing results with and without the soil erosion control project. It is at this stage that the notions of financial and economic or social prices come into the picture. The process of valuation at this stage becomes an overall mix of economic and physical values, all of which were estimated at some earlier stage.

The sixth step consists of defining a criterion for investment decisions, or what is often called a "measure of project worth". Examples of such measures are "internal rate of return" (IRR) and "net present value" (NPV). These measures of project worth provide the basis for acceptance or rejection of a soil conservation project as a whole, or of any individual part of it.

The seventh step consists of analyzing perceived project risks. A technique often used in this step is "sensitivity analysis". Sensitivity analysis focuses on two tasks: (i) define the most important variables in the technical analysis (for example, erosion rates) and in the economic analysis (for example, prices), then (ii) recompute the appraisal calculation using different values for each variable to explore the effects of changing values for each variable (values within a spectrum of probability in each case). One way to proceed is to use those values which represent the limits of economically defensible action, that is, the value of any given parameter that will make the NPV equal to zero or the IRR equal to the opportunity cost of capital. Deciding which variables contribute most to the final measure of project worth is a matter of judgement, experience and familiarity with the way in which the analysis was carried out. The magnitude of any given switching value of a variable is extremely important, both in locating and assessing the project's risks and in revising the major assumptions used in the technical and economic analysis where this is found to be appropriate or necessary. If, for example, it only takes a small drop in prices to make a project economically unacceptable, assumptions about prices must be carefully scrutinized.

An Example

What follows is an example of economic evaluation of flood damages. In general, a flood is defined as a hydrological phenomenon involving abnormally high water runoff that is not confined to normal channels. Flooding often affects land used for human activities, such as farming or housing.

The objective of the evaluation process is to assess flood damage costs "with" and "without" a given flood control project. (Most of the benefits of such a project will be associated with avoidance of losses, for example, the value of crop output saved by the project). But before one comes to the conclusion that investments in flood control are economically sound, one must go through the series of steps outlined in Section A of this chapter.

The First Step

For the sake of simplicity we assume that there is evidence that downstream flood damage would be reduced by erosion control measures.

The Second Step

This step consists of analyzing the major physical characteristics of the floods. First, one has to identify the main natural characteristics that affect the level of runoff. The characteristics most commonly considered are, size of catchment area, average rainfall, intensity and duration of rainfall, soil characteristics, soil water status (for example, moisture capacity), vegetation cover in the catchment, topography of the catchment (for example, slope) and shape of the catchment area.

The resulting flows are analyzed by means of hydrographs, which show the relationship between volume of flows and the time over which the flows change, following precipitation in the catchment. A decision must then be made about the type of flood one wants to control (for example, a peak flood occurring on average once every 50 or 100 years). This is called the "design flood".

Four characteristics of floods are often studied in detail, (1) velocity, (2) depth, (3) inundation period and (4) area inundated. These characteristics are extremely important when it comes to assessing the monetary value of losses (that is, the savings that will be associated with the flood control project). A few examples will illustrate this point. High velocity floods will cause more damage than low velocity floods, particularly to infrastructure and livestock. Depth of floods is also important in determining flood damages to agriculture and urban areas -- that is, a flood whose depth is a few centimetres will cause less damage than two or three-meter floods. Length of inundation period is a particularly critical variable in determining losses in crop production -- that is, a flood that lasts only a few hours will produce little or no damage compared to a flood that lasts several weeks. The size of the area inundated is significant because there is a very high correlation between flood damage costs and the size of the area affected.

Because of the random nature of rainfall, the predicting of floods requires an assessment of probability. A mathematical expression that summarizes the main variables involved can be expressed as:

$$Q_p = C \times R \times S \times A^{3/4}$$

Where:

Q_p = the flood peak estimate
C = the characteristics of the catchment
R = the rainfall factor
S = the catchment slope factor and
A = the catchment area

The variable C pertains to the catchment's infiltration and surface cover characteristics. The variable S is a function of the size of the catchment area and the longest distance between the point where measurement is taking place and the farthest point of the catchment.

The next stage is to define the type of flood one wants to control, "the design flood". Following an analysis of rainfall, area effects and probability or occurrence, one may construct a series of tables characterizing the potential physical damage of different kinds of floods.

The Third Step

The definition of units of account for establishing flood damage costs depends on the type of economic activity affected by any given flood. When an agricultural area is affected, a common unit of account could be the value of output foregone due to flood.

It should be clear that the total costs of floods depend not only upon the value of damaged agricultural crops but also upon other changes in the physical output lost due to floods (for example, livestock). The estimation procedures of the hydrologist and the agronomist will determine the quantity of output foregone due to any given type of flood. It is at this point that the integration of the technical framework with the economic framework becomes a key element in the overall economic assessment of the benefits of flood control projects. It should not be forgotten that the value of crops lost or damaged due to flood are only partial valuations of flood damages. There are many other losses and costs (to humans, to plant and animal life) that are not included in the analysis.

The value of the financial investment (that is, the cost of the flood control project) will depend on which type of "design flood" is used. The same applies to other variables. Similarly, the value of output that would otherwise be foregone (that is, the benefits of the flood control project) depends not only on the specific values of the commodities (or infrastructure) damaged by floods but also on how different characteristics of floods and probability of occurrence, are assessed.

Other Steps

Based on the results obtained in previous steps, the economist continues the analysis. Of particular importance in this case is sensitivity and risk analysis. Notwithstanding the importance of a sensitivity and risk analysis in relation to potential flood damages in general, the sensitivity analysis should also focus on key parameters that are thought to affect the outcome of the quantitative analysis (for example, yields and area flooded).

The results of all this analysis are then presented to those who will make the investment decision. Since there is necessarily some degree of uncertainty about the variables, it is important that the analysis be as complete as possible and that the variables with the greatest degree of uncertainty be identified. In addition, non-quantifiable effects (of floods, or of the proposed project) should also be included in the analysis. These may be important social, historical, or scientific effects that cannot be easily quantified but should not be ignored.

8

Technical Frameworks and Relationships: Erosion

Introduction

A project evaluation's ultimate objective is to assess welfare effects, in this case the effects of soil erosion control programs. An outline of the most important physical factors affecting erosion rates was presented in Chapter 2. Unless these physical elements are measured and related to their major welfare effects on different groups in society, the economic evaluation process will be meaningless. These effects should include both upstream and downstream effects.

The validity and usefulness of such economic measures as productivity, costs and profit gains and losses will be determined by the selection of technical relationships and measures. This chapter outlines and explains the analytic frameworks that can be used to determine technical parameters. Predictions of how the physical variables will behave "with" and "without" a soil conservation program are the main thrust here.

Given the large number of frameworks and variables and the many alternative ways of quantifying their behaviour, the approach here is largely descriptive. Major emphasis is on explaining the types of relationships used most frequently and in presenting a few of the quantitative frameworks most widely applied.

Since specific environmental characteristics are often the determining factors in choosing the final set of variables to be included in any technical appraisal, the presentation focuses on general approaches. The case studies presented in Chapter 11 illustrate how these approaches have been applied and modified in practice, depending on circumstances.

On-Farm Erosion Effects: Predicting Erosion Rates

The quantification of on-farm effects on productivity is a complex task. Many steps must be taken before it is possible to assess the potential effects of soil erosion on the soil's productive capacity. If one uses an input-output format to illustrate the relationships that are to be quantified, three basic steps need to be considered, (i) the role of inputs, (ii) the role of soil properties and (iii) the role of changes in output.

A distinction is made between controlled and uncontrolled inputs. Erosion is the product of both erosivity and erodibility. This section analyzes erosivity factors first and then deals with erodibility. Two important predictive models, The Universal Soil Loss Equation and CREAMS, are presented at the end of this section.

Quantification of Erosivity

As noted earlier, studies of erosivity focus on the role of rainfall in detaching soil particles. Rainfall is part of the hydrological cycle. This cycle is the continuous movement of water from the sky to the ground, through and over the ground to the ocean and back again to the sky. The evaporation that takes place above the ocean is transformed into clouds through condensation. These clouds then generate precipitation that comes to the ground, depending on temperature and other factors, precipitation may take the form of rain, sleet, fog, or snow. The water that hits the ground may be lost through runoff, percolation to a groundwater aquifer, evaporation, or transpiration via plants, trees and vegetation cover. The runoff goes to rivers, streams and lakes. Water may also be stored in the soil or deposited on the surface in soil depressions or in artificial storage structures.

Two important relationships must be understood in linking rainfall with erosion. The first is the relationship between rainfall intensity (or quantity of rain falling in a given time) and the cumulative energy of all the raindrops making up the rainstorm (which is a function of the velocity of the raindrops and their mass). The second relates the volume of runoff water over the surface of the ground to its energy for detachment and power for transport. Detachment of soil particles, either by raindrops or by flowing water, is an energy transfer process. It can be expressed rationally by a kinetic energy term. Kinetic energy, E, is expressed by:

$$E = \Sigma \tfrac{1}{2} MV^2$$

Where Σ signifies the sum of all raindrops in a storm, M is the mass of each raindrop and V is the velocity of rainfall.

Since it is impossible to measure the mass and velocity of all raindrops in a storm, intensity of rainfall has been used as a proxy measure because meteorological sites commonly have automatic recording rain gauges to make such measurements. Table 8.1 presents several methods of relating energy to intensity. The precise relationship depends on type of storm (convective, monsoon, drizzle, and the like), drop-size characteristics and methods of measurement.

Nevertheless, there is reasonably good agreement among studies in the United States, Japan, Trinidad and India about the principal determinants of kinetic energy. Kinetic energy has been shown experimentally to be a good measure of the ability of raindrops to detach soil particles.

Table 8.1 Vegetal Cover Factors (C) for Some West African Conditions

Factors	C: Representative Annual Value
Bare soil	1.0
Dense forest or culture with thick straw mulch	0.001
Savannah and grassland, ungrazed	0.01
Cover crops: late planted or with slow development	
First year	0.3-0.8
Second year	0.1
Cover crops with rapid development	0.1
Maize, sorghum, millet	0.3-0.9
Rice (intensive culture, second cycle)	0.1-0.2
Cotton, tobacco (second cycle)	0.5
Groundnuts	0.4-0.8
Cassava (first year)	0.2-0.8
Palms, coffee, cocoa, with cover crops	0.1-0.3

Source: Roose, E.J. "Use of the Universal Soil Loss Equation to Predict Erosion In West Africa" in Soil Erosion: Prediction and Control, paper to a National Conference on Soil Erosion, Soil Conservation Society of America, May 24-26, 1976, pp. 61-65. Reprinted by permission of the Soil and Water Conservation Society.

Measuring the transfer of energy by flowing water is a different problem. Both the duration of a storm and the length of time that high intensity is maintained seem to be important here. This is a vital question, since storms in the tropic tend to be far more intense and concentrated than those in temperate areas.

Erosivity indices have been derived to measure the combined energy transfer of raindrop detachment and flowing water. These indices are usually empirically derived -- that is, from experiments in which rainfall input and soil loss output has been monitored without attempts to identify how the mix of processes causes erosion. One of the most common indices is EI_{30}, which is used in the Universal Soil Loss Equation. EI_{30} is the product of the kinetic energy of a rainstorm (E) and maximum sustained intensity lasting for 30 minutes (I_{30}). Some researchers believe that this index works (one cannot say "explains", because it is empirical) because its first term expresses raindrop splash and detachment and its second term the transporting power of runoff. Note however, that EI_{30} is suitable primarily for bare soil conditions, other indices are better when there is a vegetation cover.

Because an erosivity index like EI_{30} is not normally measured directly at meteorological stations, researchers have identified relationships between total rainfall and the index. These are useful in drawing erosivity maps to show where rain has the potential ability to cause erosion (Moore, *et al.* 1979). Kingu, (1980), has quantified erosivity indices for many developing countries.

Erodibility

Erodibility is the susceptibility of a soil to erosion. Strictly speaking, it is the intrinsic properties of a soil that prevent or allow detachment and transport of particles. Central to the concept of erodibility is the soil's structure and physical attributes, especially its water infiltration characteristics, its propensity to surface crusting and sealing and the stability of its aggregates (lumps or clods) when struck by raindrops. The type and degree of vegetation cover of a soil and its past erosion history, are important determinants of a soil's erodibility.

It is important to consider the technical framework of a relationship between a soil and its susceptibility to erosion. The framework should include the physical parameters of a soil (such as particle size, sand, silt and clay). It may also include composite parameters (such as percentage of water-stable aggregates above a certain size) or measures of a soil's reaction to water (such as the degree to which clay disperses in water). Some frameworks go further and consider how a soil changes its characteristics because of types of tillage, forms of planting, mulching and amount of previous erosion. One of the simplest erodibility indices is:

$$\frac{\%\ \text{sand} + \%\ \text{silt}}{\%\ \text{clay}}$$

This merely relates the particle-size attributes of a soil. This formula does reflect the susceptibility of some soils to erosion, but it fails to bring out some of the important erodibility characteristics of most soils, particularly tropical ones.

One of the most comprehensive indices of soil erodibility is known as the K-factor, which was derived from work on a large number of soils throughout the United States. Because it was derived empirically, the K-factor is not necessarily applicable to other countries, but it is now widely used as an index. Table 8.2 gives K-factor values for several soils and environmental conditions.

Other Erosion Factors

Topographic factors are also important, slope gradient, slope length and sometimes, slope shape. The steeper the slope, the greater the splash action of raindrops and the greater the velocity of runoff water causing scouring and transport. The greater the length of slope, the greater the volume of runoff that accumulates down slope. More runoff means a higher velocity of flowing water and hence more erosion.

Table 8.2 Representative Values of the Erosion Control Factor K

USA	(Wischmeir and Smith, 1965)	K
	Contour Ploughing	0.75
	Ploughing and contour trenches	0.5
	Ploughing and contour grass strips	0.25
West Africa	(Roose, 1976, 1977)	
	Contour trenches	0.20-0.10
	Anti-erosive strips (2-4 m wide)	0.30-0.10
	Straw mulch (6 tonnes/ha)	0.01
	Curasol (60 g/1/m^2)	0.50-0.20
	Temporary grassland	0.5 -0.1

Roose, E.J. "Use of the Universal Soil Loss Equation to Predict Erosion In West Africa" in Soil Erosion: Prediction and Control, paper to a National Conference on Soil Erosion, Soil Conservation Society of America, May 24-26, 1976, pp. 61-65. Reprinted by permission of the Soil and Water Conservation Society.

Experimental research has shown the separate effects of slope length and steepness. Length is defined as the ground distance between the top of a slope and the nearest watercourse or point where spill deposition takes place. It can also be defined as the downhill distance between contour embankments. As slope length increases, erosion increases by a disproportionately greater amount. For example, a doubling of slope length can mean up to a quadrupling of rate of soil loss. The situation is similar with respect to slope steepness. As the velocity of water increases, its transporting capability may increase to the fourth power (for example, double the velocity gives 32 times the transport).

Vegetation cover is an important factor. In areas of high intensity rainfall, the intercepting capability of leaves and stems is critical in reducing erosion. The vegetation absorbs the initial impact energy of the raindrops and the drops then fall to the ground surface at a much lower velocity (and hence lower energy).

Mulch and close-growing ground vegetation are ideal protectors of the soil surface. Higher canopies, such as those provided by trees, allow drops to reform and pick up a certain proportion of their previous energy as they fall to the ground.

The relationship between erosion and vegetation cover is curvilinear. As the average cover of the ground surface (measured as the proportion of ground directly protected by leaves) increases from zero to 40% there is little added protection against erosion. But protection increases rapidly above 50%. At 70% mean cover throughout the growing season, for example, erosion is only about one-hundredth the level it would be if the soil was bare.

Vegetation also has several other effects. First, it directly reduces the velocity of runoff by impounding some of the water and releasing it slowly. Second, its roots physically bind the soil, thus stabilizing soil structure and promoting greater infiltration of water. Third, the vegetation introduces organic matter into the soil, which promotes fertility and better plant growth. This reinforcement process, in which vegetation promotes its own growth by improving its soil environment, is a key aspect of soil conservation.

Cropping systems and sequences, types of tillage and residues, harvesting methods and the like can all contribute to better vegetation cover. Any management strategy that seeks to encourage vegetative growth is, in effect, a conservation approach. This includes choice of land use, mulching, stubble incorporation, green manuring, intercropping, sequential planting and many other techniques. The precise effects are difficult to quantify, but these techniques have brought about great reductions in soil loss throughout the tropics.

Predicting Erosion Rates: The Universal Soil Loss Equation (USLE)

The factors influencing soil erosion described above have been taken into account in predicting how many tonnes of topsoil are lost in any given area. The analytic framework most commonly used has been named the Universal Soil Loss Equation (USLE). This equation was developed in the United States, using data primarily from the Midwest.

The following is a short description of the USLE. Readers interested in a comprehensive explanation should refer to the USDA handbook. The equation is:

$$A = R \times L \times S \times K \times C \times P.$$

Where, A = soil loss per unit area, R = rainfall, L = slope length, S = Slope steepness, K = Soil erodibility, C = cropping and management and P = conservation practices.

Rainfall (R). The USLE employs the index EI_{30} which is, kinetic energy times maximum sustained intensity lasting for 30 minutes. Maps of this erosivity index have been drawn for many countries and can be used to find values appropriate to a particular site.

Slope length and Steepness (L and S). These two factors are entered separately. A standard slope with a length of 22.1 meters (72.6 feet) and a steepness of 9% has been established. Tables are available which give the "topographic factor" (length and steepness combined) when the length and steepness of a slope are known. Tables are not available for slopes longer than 300 m or steeper than 20%. There is some doubt as to the accuracy of the data for slopes more than 100 m and 15%.

Soil Erodibility (K). This is the natural susceptibility of a soil to erosion. K is defined as the rate of soil loss per erosion index unit as measured from a standard plot. Most K-values vary from about 0.03 to 0.07. The K-value should be locally determined by field plot experiments, but a nomograph developed in the United States that shows how a combination of soil parameters are affected by erosion is often used. Extreme care must be exercised when it is used for soils in countries other than the United States.

Cropping and Management (C). This is defined as the ratio of soil loss from land cropped at a particular site to the corresponding loss from a predetermined standard (clean-tilled, continuous fallow). The ratio is intended to reflect the combined influences of vegetation cover, type of crops, tillage and management techniques. C-values are determined experimentally and have been presented in tabular form for a large number of cropping conditions prevalent in the United States.

Conservation Practices (P). P takes into account the effectiveness of erosion control practices, such as contouring, terracing and strip-cropping. As with C, it is a ratio comparing soil loss at a particular site to soil loss under a set standard. Values of P have to be experimentally determined.

The USLE is universal only insofar as the principles by which it operates are universal. A distinction must be made between the equation and the values entered into the equation. Merely to transfer the values derived in the United States to other climates, topographies and farming conditions would be a grave error. Furthermore, the USLE is applicable only to sheet and rill erosion. That is, it is not applicable to gully erosion. Modifications of the USLE have been proposed to predict wind erosion and erosion under other environmental conditions, but it should be recognized that these modifications have not been fully evaluated.

The USLE was designed for the following uses, (1) to predict annual soil losses from fields (it was not designed to predict soil losses in watersheds, building sites, regions, countries and the like), (2) to guide the selection of conservation practices for specific sites, (3) to estimate the reduction in soil loss attainable from various changes that a farmer might make in his cropping system or cultivation practices, (4) to determine how much more intensively a given field could be safely cropped if it were contoured, terraced, or strip-cropped, (5) to assess the design and application of other conservation measures on farmers' fields.

Limitations of the USLE

Three types of criticism of the USLE have been voiced: technical, practical and economic.[1] From a technical viewpoint, the USLE has the following weaknesses. First, it was designed only to predict soil loss from sheet and rill erosion and therefore underestimates erosion of other kinds. Second, not all sediments leave a particular field and the equation was not designed to predict sediment yields outside the field. Third, it only predicts average soil losses and more work needs to be done to estimate loss "variability". Fourth, since the equation is empirical and derived from specific conditions, researchers should be aware of the problems of extrapolating it to other situations. Fifth, there are several potential sources of error: Using factors with too broad a base, extrapolating factor relationships beyond data capabilities and defining slope length factor incorrectly. Sixth, the equation is not very useful for determining soil erosion from irregular slopes. Seventh, the equation assumes that the factors in erosion are independent. Since they are not, this can lead to a failure to perceive important interactions between factors.

In order to apply the USLE to a new area, it is necessary to determine the factor values for the equation, preferably from local experiments. The derivation of factor value however, is expensive in terms of effort, time and resources. It would be impossible for developing countries to emulate the research effort carried out in the United States. Certain shortcuts may be feasible, but the inherent possibility of error should be understood.

The main problem from an economic point of view is the absence of a linkage between soil loss estimates and farm productivity. Whatever the technical merits of alternative cropping practices, the adoption of any such practices will depend mainly on economic factors.

Some Comments

Clearly, measurements of losses in topsoil need to be related to land productivity and eventually to a notion or index of welfare. The quantification made possible by use of the USLE is only one element in decision-making.

One simple way to start relating soil losses to economic gains or losses, is by defining a *tolerance level*. This tolerance (T) level based on expert judgments, is a relatively subjective matter. The T level is defined as the maximum level of soil erosion that will permit a "high level" of crop productivity to be maintained economically and indefinitely. For the United States, a soil loss tolerance ranging from 5 to 10 tons/acre/year has been defined as "normal". Estimates of the rate of soil formation are an important factor in establishing levels of soil tolerance. This value will vary, depending upon soil type. Areas with a deep root zone may exceed the 5 ton per acre tolerance level without any significant loss in productivity, while shallow soils may have a safe tolerance level of less than 2 ton per acre.

The use of the T-level has major advantages due to its simplicity, its modest data requirements and the fact that it is a standard measure that can be used for both soil loss and sedimentation. But the T-level has many problems[2]. First, it is a one-value variable that does not measure relative intensity, it is not indicative of how bad the erosion problem would be if topsoil losses were 10 tons/acre/year instead of 20 or more tons/acre/year. Second, the concept is also limited in that it is not directly related to any notion of productivity gain or loss. There is thus no economic content in the index, making it very difficult to apply to economic decisions.

A simple but useful formula to show intensity can be formed by computing the difference between the estimated amount of soil losses and the T level, divided by the T level:

$$\frac{A - T}{T}$$

This formula will produce a series of values which can then be related to soil productivity.

To approximate a relationship that would be more meaningful in determining gains and losses in productivity, several other elements need to be brought into the analysis. One is soil depth. While most of the discussion has focused on the soil that leaves the farm, it is also important to study the characteristics of the soil left on the farm. Two important characteristics should be mentioned, soil depth (or root zone depth) and nutrient availability. These are often used to relate erosion to losses in productivity after incorporating such other factors as weather conditions, soil moisture, water hydrology, sedimentation and chemical use.

It is not easy to outline the critical steps in evaluating the effects of soil erosion on yields. There are no standard techniques for such an evaluation. Examples of how some of these relationships have been analyzed however, may prove useful to those planning to appraise soil conservation projects.

Yield-Topsoil Relationship

One useful way to estimate the impact of erosion on yields is by relating erosion to topsoil depth. The main assumption here is that, if all other factors are held constant, yields will decline as more topsoil is lost (Young and Walker, 1982).

This relationship has been estimated through experiments on crop response to different levels of topsoil. In mathematical terms, the relationship can be expressed as follows:

$$Y^t_s = f(D,Q,M,W,t)$$

Where Y^t_s = crop yield per acre in time t, D = depth of topsoil (in inches), Q = soil chemistry factor, M = management factors, W = weather or other climatic factors, t = time, f = functional relationship.

Obtaining estimates of these yield functions requires many observations under different conditions. These estimates are usually nonlinear and are specific to the type of soil and climate under investigation.

Technology-Yield Relationship

Advances in agricultural technology are crucial to the control of erosion. For example, the development of high-yielding varieties or crops that fixed nitrogen more effectively could potentially reduce erosion by providing more plant cover. Because of technological advances, one often finds situations where yields continue to increase despite continuing erosion.

Nutrient Budgets

Economists view productivity as the relationship between inputs and outputs. Soil nutrients (for example, nitrogen, phosphorus, potassium) have a major impact on plant growth. It is often said that there is a higher concentration of nutrients in topsoil than in subsoil. Losses in nutrients are directly correlated with losses in productivity, although compensating factors may exist. Nutrient availability is important in determining soil quality and therefore, productivity (Russell, 1980).

Experiments have been carried out to determine the effects of nutrient availability on plant growth or yields. These experiments have focused on the effects of one nutrient or the interaction of several nutrients. These studies have supported the intuitive notion of diminishing returns, plant growth will increase more slowly as more nutrients are supplied until it reaches a point where no net growth takes place. If very large quantities of certain nutrients are supplied, plant growth may even decrease. A mathematical expression often used to predict yield changes due to variations in nutrient availability is the following:

$$Y = A\,(1 - e^{-cx})$$

Where, Y is the yield obtained, A is the maximum yield obtainable, c is a Constant, and x is the quantity of nutrients.

An understanding of this relationship is extremely important in the context of soil conservation programs. As nutrients are lost due to erosion, plant stress becomes more acute and thus has greater effects on yields. In sum, two points should be stressed. First, in assessing productivity, the quantity and quality of existing soil is what really matters, second, the technical measures presented up to this point reveal the need to develop an expanded framework.

The presentation above shows clearly that predicting the effects of erosion on productivity is not a straightforward matter. The number of variables that influence change in crop yields is very large. Until there is a more precise way to quantify this relationship, little progress will be made in improving the technical and economic evaluations of soil conservation projects.

This section presents the main characteristics of a productivity model developed by Pierce, *et al.* (1983). The first and perhaps most challenging, question is to define soil productivity. In general, soil productivity is defined in terms of its capacity to produce a crop or a sequence of crops under predefined physical and management conditions. Highly productive soils are characterized as having a relatively high volume of organic matter, medium texture, good tilth and a depth of 150 cm or more.

The central analytical question, therefore, is to determine the complex relationship between plant growth and soil attributes. The Productivity Index model quantifies soil productivity by means of a numerical index method that relies on five soil characteristics, moisture capacity, bulk density, aeration, pH and electric conductivity. Each of these characteristics was evaluated to determine their impact on root response and each soil layer was weighted based on an "ideal" distribution of rooting. The index is normalized to take values between 0.0 and 1.0.

The Productivity Index takes the following form:

$$PI = \Sigma^{n}_{i=1} (Ai, Bi...., WFi)$$

Where: A is the sufficiency of available water capacity, B is the sufficiency of aeration, C is the sufficiency of bulk density, D is the sufficiency of pH, E is the sufficiency of electrical conductivity, WF is a weighting factor, PI is the productivity of the soil environment, r is the number of horizons in rooting zone under ideal conditions, i is the increments of soil depth.

This model has been modified to the following:

$$PI = \Sigma^{n}_{i=1}(Ai, ..., WFi)$$

Considerable research has gone into determining the correct values for these variables. An interesting evaluation of its use in developing countries has been published by Rijsberman and Wolman (1984). Attempts to apply the Productivity Index to tropical soils are complicated by the particular attributes of those soils and the very limited information on how productivity is affected by each factor individually.

Computer Models for Erosion Prediction

Several models have been developed to estimate and simulate changes in erosion rates. One important model, frequently used in the United States, is CREAMS, a field-scale model for Chemical Runoff and Erosion from Agricultural Management Systems[3]. The presentation here focuses on the main relationships underlying erosion analysis without elaborating in depth on the mechanical and mathematical aspects of this model.

The main feature of CREAMS is that it predicts erosion rates in terms of sediment leaving a given field (farm). It also provides estimates of surface water hydrology, sedimentation, nutrient losses and chemical composition. There are two types of inputs, natural inputs and management inputs. The most important natural inputs are precipitation, solar radiation and temperature. The most important management inputs are land use, plant nutrients, cultivation practices and pesticide use. Depending on the geology, soil characteristics and topographic factors of watersheds, the model can be used to estimate rates of evapotranspiration, runoff and percolation. On the basis of surface runoff estimates, the model can be used to quantify erosion, sedimentation and chemical changes.

The model has several advantages. It can integrate several types of well-documented technical frameworks, it uses computer codes that are easily available, it is relatively cheap to run and it has been applied in a number of places of the world. Its major disadvantages are that it requires much more data than the USLE, must be worked out by computer and is more appropriate for single fields than for watersheds.

Computer Models to Predict Yields

Computer models have also been developed to translate the effects of erosion into yield estimates. One important model is EPIC: Erosion Productivity Impact Calculator. EPIC has seven submodels, hydrology-weather, erosion, tillage, nutrients, plant growth, soil temperature and economics.

The soil properties, water and nutrient submodels are intended to simulate factors that affect the levels of resources available for plant growth. The weather simulator denotes the effects of such factors as rainfall and air temperature. The plant growth simulator includes the effects of solar radiation on the plant maturity process, thermal time and the initial biomass of nutrients and residues.

The tillage and residue management portions of the model simulate the effect of cultivation practices as it relates to erosion and thereby determines, for example, how tillage practices affect productivity and erosion rates. The soil erosion submodel uses the USLE or the Modified USLE (MUSLE).[4] As with most other models, EPIC needs validation and calibration. It may eventually be useful in developing countries if sufficient data become available.

Notes

1. See Wischmeier (1976), where the author of the USLE carefully specifies the limitations of the equation. Many unwise and inappropriate applications of the equation have been made.

2. In addition, one needs to consider the fact that whatever small value one uses, long term soil losses will create rooting problems and, thus, long-term decreases in yields. On the other hand, a very low level for T may be achieved only at a very high economic cost.

3. USDA, Conservation Manual Report No. 36.

4. A detailed description of this model is found in J. R. Williams, *et al.* (1982).

9

Technical Frameworks and Relationships: Downstream Effects

Quantifying the Effects of Sedimentation

In general, analysis of sedimentation involves the characterization of at least the following aspects, (1) the catchment area, (2) the movement of sediments, (3) the physical attributes of water channels and (4) the nature of deposition. With regard to the characteristics of the catchment area, models have been designed to simulate various physical aspects (for example, topography, geology, vegetation, land use) and hydrological aspects (for example, rainfall/temperature, movement of rainstorms, snow accumulation, evaporation) and their interactions.

Analysis of sediment transport and deposition also requires a knowledge of the physical characteristics of sediments (for example, size, diameter, shape and density of particles), fall velocity, threshold movements, bed load, suspended bed load material, fine material or so-called washload and river or channel bed features.

Many hydrological models analyze channel geometry and the factors affecting such geometry.

Another group of models assesses the mechanical aspects of sediment transport in various degrees of complexity. Consequently, one finds models to assess sediment transport and deposition in lakes and water reservoirs, flood plains and estuaries. As an integral part of these models, hydrologists also focus on the efficiency of different types of structures to "trap" these sediments and on the dynamic aspects of sediment movements in a given reservoir area.

An important use of such models is to help in designing the water reservoir. A distinction is often made between the "dead" area and the "live" area of a reservoir. The dead area is that part of the reservoir that is expected to be filled (totally or in part) with sediments. If sedimentation rates are higher than predicted during the design stage, the volume of sediments deposited will compete for space with water, decreasing the live area of the reservoir.

Before any project is designed, efforts are made to predict the rates of sedimentation and deposition. UNESCO (1982) has classified alternative methods of prediction into three categories, (1) empirical methods, (2) calculation methods and (3) modelling methods.

Empirical methods focus on sediment yields from watercourses. This is done by way of collection and analysis of data on suspended and bed load samples, or by means of surveys of the volume and nature of sediments transported into impoundments, reservoirs, or deltas. These methods require the collection of data relating to catchment characteristics, since the main source of sediment formation is erosion. *Calculation* methods focus first on the erosion process. These methods then address the hydrodynamics of sediment transport and the properties of sediment particles, the main characteristics of the water course (for example, width, depth, slope) and the stability of water channels. *Modelling* methods use simulation techniques to characterize some of the previously outlined factors as well as the dynamic nature of water flow, transport of sediment and movement of pollutants. Advances in computer hardware and software have facilitated the use of modelling methods.

Not all the sediments moving from their original location enter streams. Some are deposited on other agricultural or non-agricultural lands before they reach a stream. The erosion-sediment process starts with the detachment of sediment particles through mixing with water, which acts as a transport agent and then with their deposition and subsequent compaction elsewhere.

The amount of sediment generated depends on several factors, the most important being the watershed factor (for example, ground cover, soil type, water flows and crop management). Measurement of sediment yields can either be done experimentally or with predictive models. Experiments are usually based on the sampling of sediments in fields, stream flow sediment analysis, monitoring of sediment load in flood waters and surveying the bed profile of reservoirs. What is accumulated in a reservoir will depend on the efficiency of that reservoir in trapping sediments.

Predictive models that have been developed include models to determine sediment transport functions, upland erosion models (USLE and MUSLE), sediment delivery ratios (the fraction of sediment runoff that passes any particular point), stream bank erosion and watershed management.

Sediment-Yield Production. Glymph (1975), notes that there are four methods for predicting sediment yields, (1) the sediment-rating curve flow-duration method (SRCF), (2) the reservoir sediment deposition survey method (RSD), (3) the sediment delivery ratio (SDR) method, and (4) the bed load function method (BLF). The SRCF requires measurement of stream flow and amount of sediment. This method is used to predict how many tonnes of sediments are expected to enter reservoirs, tributaries and main rivers. This method is not useful in estimating the total sediment production of watersheds. The RSD method focuses on the measurement of sediments deposited in reservoirs or sediment ponds. The volumes are converted into weight units, modified by the capacity of any reservoir to trap such sediments. The SDR method focuses on the relationship between the sediment yields produced by a watershed and gross erosion in that watershed at any point in time. The BLF method provides a framework for quantifying the sediment rate (that is, the proportion of sediment in the water) and quantity of materials arriving in the beds of alluvial channels. The information needed to use this method includes particle size, gradients and gross sections and flow duration.

The sediment delivery ratio (SDR) varies, depending on the magnitude and proximity of the soil to the stream flow, the transport system, the size of eroded material and such non-watershed characteristics as the size of the drainage area and its topography. Suspended sediment loads are estimates of average sediment concentration in relation to stream flows. By means of conversion factors, figures on tonnes per day of sediment can be obtained (Holeman, 1968). Sediment yield curves have been obtained from empirical observation. These indicate the relationship between the size of the drainage area of any given watershed and the corresponding sediment load. As a drainage area gets larger, sediment yield ordinarily becomes smaller. As a drainage area gets larger, sediment yield *per unit area* also gets smaller.

Trap Efficiency of Reservoirs

The trap efficiency of a reservoir is often defined as that percentage of the total inflow of sediments that is actually deposited in the reservoir, given the type of outlets and the operating rules of the reservoir[1]. In general, a small reservoir will not be as efficient as a larger one in trapping sediments.

The potentially negative effects of increasing the sedimentation rate in reservoirs will be determined by the path, location and process of sediment deposition. Mathematical models have been developed to quantify these factors. Intuitively, one would expect that large particles will settle out first and the finer silts and clays last.

Computerized models are now in use to estimate the amounts, rates and spatial distribution of sediments in lakes and reservoirs. One of these models is SEDRES, which computes sediment entrapment and differential settling for three classes of sediments (that is, clay, silt, sand) and their compaction. Inputs to the model are water inflows, reservoir operation levels, reservoir location, its elevation in the watershed, its area and capacity, sedimentation characteristics, type of sediment entrapment and sediment distribution method.

Water Quality Changes

There is great concern about changes in water quality due to erosion, even though erosion is only one factor in water quality. Sediments carry nutrients, chemicals (pesticides) and soil particles, all of which can cause severe damage downstream. Excessive concentrations of nutrients, for example, increase the growth rate of weeds in reservoirs, which in turn decreases reservoir storage capacity. Concentrations of harmful chemicals decrease the economic life of hydroelectric equipment and yields from downstream irrigation, increase the costs of water treatment for drinking purposes and damage wildlife and fisheries habitats. Excessive turbidity and sediment deposits impair navigation and recreation.

The analytical frameworks developed to determine changes in water quality vary, depending on their purpose. It is therefore more difficult to present clear-cut frameworks for this purpose than for measures of sediment volume. Here, we attempt to identify technical relationships that can be useful in project analysis.

It is important to note however, that a major part of the necessary analytical work in this area is based on erosion and sediment transport models presented earlier. Nevertheless, it is important to briefly point out other aspects of water quality analysis.

The subject here is measurement of changes in the quality of water (vis-a-vis water quality standards) caused by leaching. Several models have been developed to assess the effects of water quality on the economic life of hydroelectric equipment (for example, turbines) and on facilities for treating water later used for irrigation or drinking.

Pesticides and fertilizers which are beneficial in increasing crop productivity leach into the ground and eventually reach water aquifers and rivers. Concentrations of these chemicals in different parts of the hydrological system affect the quality of water, which in turn is an input to other economic activities or other ecosystems (for example, fisheries and wildlife). The optimal level of chemicals or nutrients in water varies a great deal depending on water use. A nutrient level that is acceptable if the water is to be used for irrigation may be unacceptable if the water is to be used for drinking. By the same token, water acceptable for drinking may be unacceptable for ecosystems such as fisheries.

The extent to which water quality is affected by leaching or runoff, depends on soil type and the solubility of certain chemicals. Phosphates, for example, are less soluble than nitrogen. Chemical processes also affect the rate of solubility. Models that take chemical composition into account have therefore been developed.

Turbidity Measures. Turbidity measures depend on both sediment concentration and the characteristics of sediment particles. A given concentration of small soil particles will result in greater turbidity than the same concentration by weight of larger soil particles. This is due to the increase in surface area of the smaller particles per unit of weight. Since erosion control practices have differential impacts on the particle size of sediments, it is not easy to relate changes in turbidity to conditions with and without given project practices.

Water Pollutants. Nutrients in soil eroded from farmlands affect water quality. Nitrogen fertilizer compounds, for example, are highly soluble. Thus, associations have been made between expansion of fertilizer use and amount of nitrates in water. Pesticides also affect water quality and soil quality as well.

Notes

1. US Corps of Engineers. *Sedimentation Manual* (Draft version). November 1982.

10

Theoretical Economic Considerations

An Overview

There are three principal problems in the formulation and appraisal of soil conservation projects, the proper identification of benefits and costs, the assignment of monetary values to benefits and costs and the issue of how to make use of discounted cash flow procedures. By analyzing the erosion phenomenon in physical terms, assessing the dimensions of erosion effects, developing a typology to account for these effects and defining soils as a natural resource, Chapters 1 through 6 of this book should enable the analyst to identify the most important benefits and costs of soil conservation programs. It is very important to realize that a failure to identify benefits and costs will result in poorly designed projects. This is despite the difficulties in assigning monetary values to these benefits and costs. This may be illustrated as follows.

Suppose that a project is designed and implemented solely for the purpose of reducing the erosion rate to prevent declines in soil productivity, even though the erosion rate may be sufficiently low to prevent further decline in productivity upstream. Such a project may cause considerable damage downstream. The volume of downstream sedimentation may still be sufficient to harm such economic activities as the generation of hydroelectric power as well as flood control. Or suppose that the concentration of chemicals in the sediments is very high. In such cases, even if sedimentation rates are considered to be low, financial resources will be needed to make the water adequate for drinking and industrial purposes. Thus, even if analysts have difficulties in assessing the other two problems later on, the proper identification of costs and benefits will produce a better project design.

It bears repeating that an economic analysis of a soil conservation project is really no different, in principle, from an economic analysis of other types of projects. There is the usual mix of inputs and outputs. The two main problems in economic analysis of soil conservation projects are the *location* and *time* of effects and *valuation* of these effects. The proper valuation of these tasks is the more challenging task because a number of them are not normally traded and therefore do not have easily observed market prices. Assigning prices to nontraded goods then becomes a complex and cumbersome task.

Assuming that most of the benefits and costs of a project have been identified, the next step is to make a monetary valuation of benefits and costs. Although it is impossible to provide solutions to all the types of problems that may be encountered in the valuation process, this book outlines the most important elements.

Monetary valuation of costs and benefits is a major concern for agencies required to justify environmental projects on the basis of such criteria as internal rate of return and net present value. Since traditional discounting procedures only take into account benefits and costs which can be valued in monetary terms, soil conservation projects are often rejected because they show a low rate of monetary return. More often than not, the market assigns values only to traditional inputs (for example, capital, labor) and outputs (for example, food). Thus, the "true" social worth of a soil conservation project is often undervalued by the market because outputs, in particular, cannot be evaluated in traditional terms.

As in any discipline devoted to analysis of potential change in the environment, there will always be components or effects that will be very difficult to quantify, either in physical or monetary terms. Much more research on the quantification of environmental changes is needed. Nothing however, should prevent the project analyst from listing currently unquantifiable components or effects and presenting them to decision makers along with an analysis of benefits and costs couched in traditional terms.

Valuation of benefits and costs is not simple, even though one can always assign values to options by using either objective or subjective notions of value. The central issue here however, is to define the extent to which any process of valuation follows certain rules and to what extent those rules are widely accepted. If those rules are not followed, an assessment based on assigned values will not mean much. This does not imply that investment decisions should be based only on economic criteria. The important thing is to make the criteria that are used very explicit, so that the policy maker has a clear understanding of the origins of these values.

The main body of thought on the basic rules of monetary valuation of benefits and costs is often referred to as "welfare economics". Welfare economics is concerned with changes in the total amount of goods and services available to society as a whole, not just to individuals. This body of theory provides the basic framework for defining the *economic* criterion, (for example, benefit-cost analysis) and gives the elements necessary to determine to what extent any particular monetary value can be used in decision making. An example here may prove useful. It is often argued that sedimentation increases the cost of dredging ports. Since dredging costs are easy to quantify, any reduction in dredging brought about by a reduction in sedimentation is referred to as a "cost saving benefit" of a soil conservation project. The main question is the extent to which the monetary value of these dredging costs represents a true notion of welfare and therefore can be used in benefit-cost analysis.

We already know that detached soil and soil transported as sediments are not traded in the market. To get at some notion of a market valuation of these environmental inputs, we need to disaggregate environmental goods into their characteristics and then to relate changes in the characteristics to the production process (or production function) of commodities which are actually traded. The value of changes in environmental inputs can then be traced by assessing changes in the value of traded commodities to which the market assigns specific values. After this is done, the analyst must determine if the value complies with the basic theory of welfare economics. If it does not, modifications must be made. These modifications are well-known to most economists (for example, shadow pricing labor, pricing commodities in the presence of monopoly and shadow exchange rate).

Let us give a few examples. The first example pertains to the valuation process when sedimentation affects the quality of water used for downstream irrigation. Sedimentation as an environmental input has to be broken down into its relevant characteristics, one of which is chemical content (for example, levels of pesticides and salts). These characteristics, measured in physical terms, determine the quality of the water that will be used for irrigation. Each level of chemical concentration can be associated with a different level of water quality and water quality, in turn, can be related to crop productivity. Higher levels of chemicals will ordinarily be associated with smaller crop yields. These changes in the productivity of a tradeable commodity (for example, wheat or rice), can be given a market value through the price mechanism. Thus, the analyst can assign a monetary value to change in sedimentation rates through changes in the market value of those tradeable commodities.

Another example is the change in the quality of drinking water that results from high rates of sedimentation. Many analysts suggest that changes in water treatment costs reflect the value (or benefit) of controlling sedimentation (other things being equal). If the demand for water is given, this value will represent a true notion of welfare, since changes in sedimentation rate will affect the quality of water and therefore the cost (total or marginal) of producing drinking water.

Erosion may change the levels of nutrients in soil. As explained earlier, topsoil that is lost usually carries with it valuable nutrients. To achieve the same level of on-farm productivity, farmers must then replace these nutrients. Since we already know that topsoil is not often traded, the process of valuing topsoil may be carried out by determining the replacement value of nutrients. As is shown later on, the value of nutrient inputs, under certain conditions, is also a true notion of welfare.

This book uses "duality theory" as the main analytical framework for the "tracing" process (for example, from sediments to changes in the value of water). Since the tracing process involves several steps before valuation is possible, "duality theory" allows us to make certain simplifications. What this means is that if certain conditions are met, one can directly observe environmental protection values by observing changes in net income or profits. This will enable analysts to avoid tracing changes that take place via the production function of tradeable commodities. For example, it will not be necessary to estimate or assess how different levels of erosion affect the physical relationship of all those inputs that determine soil fertility. In other words, one would not necessarily need to follow the computations at each stage of the tracing process (for example, changes in environmental attributes, changes in the production function, changes in the profit function) if the analyst has a good representation of the profit function.

The reader who is not prepared for the analytical aspects of duality theory should at least try to understand what is involved. The reader however, does not need to understand duality theory in order to follow the presentation of evaluation methods as it unfolds here.

As stated earlier, even when the analyst has identified most of the benefits and costs of a project and has tried to value them, there may be circumstances where the rate of return may still not be high enough to justify the project. One possible reason for this is the distribution of benefits and costs when discounting procedures are applied. the main costs of soil conservation projects occur in the first few years of the project, while the benefits accrue far in the future. Because discounted cash flow techniques are in fact exponential functions (for example, $(1+r)$, $(1+r)^2$, $(1+r)^3$, ..., $(1+r)^n$) the discounted future benefits will not make an important contribution to improving the rate of return or the net present value.

Basic Principles

The assignment of values has considerable significance and generates much controversy. This is true for the assignment of monetary values or other forms of relative value. Estimates of the values of benefits or costs are necessary in order to rank investment alternatives or for accepting or rejecting investments on the basis of a specific criterion. Ranking of alternatives is necessary because demand for projects exceeds funds available. The question is not whether Project A is desirable, but what the loss to society will be if Project A is funded but Projects B and C are not.

A basic step in assigning values is to establish the major determinant of these values. The most commonly known valuation process is by way of market prices, the point at which supply and demand are equated. But the valuation of many services and commodities does not necessarily take place through markets.

Supply and demand as the major determinants of prices and exchange constitute the most common economic framework for valuation. It is important to decide when the values generated through markets provide a useful framework. It is often said that the market fails to provide meaningful price signals for activities related to the environment. Thus the sum of the marketplace behaviour of individuals may result in outcomes that are undesirable from a social viewpoint -- for example, high levels of pollution. Consequently, the selection of other methods of valuation is important. The selection process should depend on specific analytical objectives and will often be related to data availability. Different methods of valuation will require different types of data.

Another important consideration is to determine when a particular value should be used as a measure of welfare or be used to assess a change in welfare. In principle, one can assign any monetary value to a particularly commodity. But this does not mean that every valuation method will be useful in decision making. Here, we attempt to answer the following questions: First, what is the basis of valuation methods in the context of welfare economics? Second, what valuation methods are available for appraising soil conservation programs properly? And third, what are the major assumptions of each of the valuation methods proposed?

The main goal of economic analysis of projects is to select projects that are profitable to society. The notion of social profitability however, is anything but clear. Society involves many individuals and institutions and has many different objectives.

Several criteria have been proposed for judging social profitability. One of these is the so called *Pareto* criterion. This criterion states that a project is socially profitable when the welfare of every member of society is improved and nobody is made worse-off. The Pareto criterion is rarely met, since it is difficult to find projects that benefit everyone and harm no one. As a consequence, other criteria have been proposed.

The most widely accepted are the *Kaldor-Hicks* and the *Skitovsky compensation* criteria. Kaldor-Hicks states that a project will be socially profitable as long as those who gain from the project receive more than is sacrificed by those who lose, the gainers could then *in principle* compensate the losers and still come out ahead, thereby improving the welfare of all. Skitovsky states that a project will be profitable when the losers cannot "bribe" the gainers into not undertaking the project. The main difference between these two criteria and the Pareto criterion is that the former do not impose differential weights for individuals, whereas the Pareto criterion holds that a project is not socially profitable even if only one individual is made worse off. This gives added weight to the views of the one person made worse off.

Today, compensating and equivalent welfare gains (or losses), have become commonly accepted welfare measures. These gains and losses can be approximated through estimation of what economists call consumers' and producers' surpluses.

Consumers' and producers' surpluses are measures which represent the gains that accrue as a result of consuming or producing goods. Consumers often obtain goods at a price less than the maximum they would be willing to pay and producers sell goods at a price above the minimum price at which they would have been willing to supply the goods. Figure 10.1 shows consumers' surplus (CS) and producers' surplus (PS) in a one-commodity market under equilibrium conditions.

Marshall defined consumers' surplus as "the excess of the price which he [the consumer] would be willing to pay for the thing rather than go without it, over that which he actually does pay" (Currie, Murphy and Schmitz, 1971, page 124). Why is consumers' surplus a measure of welfare? Any individual derives utility from consuming a good. Utility is impossible to measure directly, but in paying for goods, individuals reveal the value, or utility, they place on that good. Total utility is money expenditure (PxQ in Figure 10.1) plus CS. The individual's demand schedule is a series of price-quantity points, each of which represents the marginal utility to any individual, resulting from the consumption of a given quantity of a particular product. This concept could be generalized to many consumers, assuming that the welfare of individuals could be summed without differentiating among the individuals to whom the welfare change accrues.

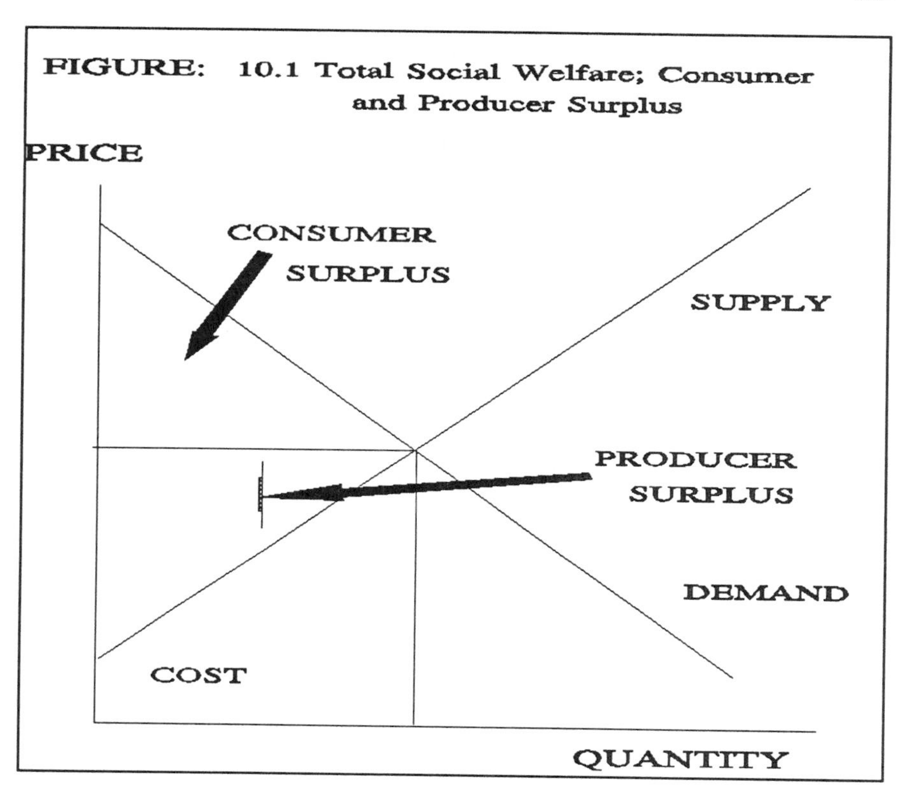

As shown in Figure 10.1, consumers' surplus is the area above the price line (P*) and below the demand curve (DD). The correct estimation of CS is very important for environmental projects. Frequently, environmental goods are supplied free -- for example, clean air, clear river water -- or at minimal cost (for example, the entrance fee to a national park). Therefore, environmental "goods" have no meaningful market price, even though consumers and society receive large benefits from them. The measure of utility (or welfare) of society's consumption of many environmental goods is therefore almost entirely CS.

The producers' surplus is the minimum price at which the producer is "induced" to supply a product. This is given, in theory, by a supply curve which represents, in effect, the value of the product to the seller. However, producers' surplus is not as readily accepted in a welfare context as consumers' surplus. To overcome the difficulties of the concept of PS as a measure of welfare, the concept of "rent" has been introduced (Mishan 1977a).

Welfare Measures for Soil Erosion Control Projects

Consumer -- Producer Surplus

In developing measures of welfare, the analyst concerned with environmental projects in general -- and soil erosion control projects in particular -- is faced with some special problems. In most cases, consumers' and producers' surplus cannot be measured directly, either by direct demand for the project itself or by demand for the potential results of such projects -- a demand for soil or for a higher level or water quality.

Many soil conservation projects have been evaluated at the micro-economic level by means of a consumer/producer surplus approach, on the assumption that the market somehow will reflect the nature of the erosion problem. The assumption is that in the absence of a soil conservation program the supply curve for commodities would shift upward due to a decrease in land productivity. Producers may gain, but the gain may be outweighed by an increase in production costs (land as a fixed factor).

In the presence of a soil conservation program where additional costs of production are incurred, gains may result from a production increase. This, in turn, may be outweighed by a decrease in price (the supply curve shifts downwards). In the absence of a soil conservation program, consumers may lose due to a decrease in supply. In the presence of such a program, consumers may gain due to an increase in the supply. Time will play an important role, as will be discussed later (Seitz, *et al.* 1979).

Some environmental outcomes (for example, soil erosion) do lead to direct measurements -- for example, increased or decreased output of products with a market value. In most cases however, one is required to value the welfare impacts of a project indirectly through related markets and proxy values. There are many cases where changes in soil quality are not reflected in the market. The main reason for this is that environmental attributes are rarely traded within markets. To solve this problem, *one must regard environmental attributes as factors of production.*[1]

To assess the welfare effects of projects, therefore, a valuation of changes in their levels should be carried out by focusing on related markets. In this case, once the difference in land productivity resulting from a conservation project has been determined, valuation is derived from the value of the crops associated with the difference in productivity.

Duality Theory[2]

The development of duality theory on production, profit and cost functions has opened new avenues for extending traditional benefit-cost analysis. Unfortunately, most of the literature on duality theory is in mathematical language that is often very difficult to grasp. Here, the presentation is straightforward, using a few concepts from basic calculus. No proof of theorems or of specific relationships is given.

As explained earlier, a key step in extending benefit-cost analysis is conceptualizing environmental characteristics and defining technical and economic relationships that will enable the appraisal team to determine how changes in these characteristics affect the productivity of tradeable commodities. In most instances, these relationships can be identified by way of the production function of commodities (for example, production of drinking water supply depends on the concentration of chemicals in one or more units of sediments), where each of the characteristics (for example, volume, size and chemical concentration) represents an input in these functions (that is environmental inputs).

Duality theory enables us to carry out an *analysis of productivity effects* by means of income, profit, or cost functions rather than going via the estimation of production functions. Under certain conditions (for example, convexity), the information that can be obtained by way of the profit function is as good as the information obtained by way of the production function. This is similar to the formulation of a linear programming problem, where one analyzes changes by way of the "primal" or the "dual".

For any practical purpose, cost function (for example, replacement cost, mitigation cost) can be regarded as a special case of profit function where output is given (for example, amount of energy use, amount of drinking water consumed).

The basic premise underlying duality theory is very simple. If farmers are behaving rationally and responding to their economic and physical environment, the analyst can learn a great deal about allocation of resources by observing the behaviour of farmers and farm-related prices. This, in turn, makes it possible to place a value on many factors, including nontraded environmental factors like soil.

Assume that a farm is producing commodity Y (for example, wheat) with a technology that uses both a nonenvironmental input X (for example, labor) and an environmental input e (for example, soil). Furthermore, assume that the price of X, equal to W, is known and that either the price of Y, equal to P, is known, or the level of output (that is $Y=Y^*$) is known.

The production function of wheat can be written as,

$$Y = F(X,e) \tag{1}$$

and the profit function as,

$$R = PY - WX, \text{ or } R(P,W,e) \tag{2}$$

The rational farmer will try to maximize profits subject to the production function, or,

$$\text{Max } R(P,W,e) = PY - WX \tag{3}$$
$$\text{s.t. } Y = F(X,e) \tag{4}$$

When the level of output is fixed (thus, $Y=Y^*$), the problem described by equations (3) and (4) can be rewritten to express a cost minimization problem,

$$\text{Min } C(Y^*,W,e) = WX \tag{5}$$
$$\text{s.t. } Y^* = F(X,e) \tag{6}$$

Based on the above formulation, one can determine the optimal use of any input (that is estimate an input demand function), an optimal profit level and an optimal output level. In particular, changes in net income due to changes in environmental quality are found by differentiating R with respect to e, thus,

$$\frac{dR(P,W,e)}{de} \tag{7}$$

To determine changes in cost (in this case, marginal cost) due to changes in environmental attributes, one can estimate the following expression,

$$\frac{dC(Y^*,W,e)}{de} \tag{8}$$

The demand function for a factor of production will be estimated as follows,

$$\frac{-\,dR(P,W,e)}{dW} \tag{9}$$

And the effect of change in environmental quality can be found by first discovering the reverse demand function for factor use, that is,

$$W = g(X,P,e) \tag{10}$$

$$\frac{dg(X,P,e)}{de} \qquad (11)$$

Finally, changes in output can be found by differentiating profit functions with respect to output price,

$$\frac{dR(P,W,e)}{dP} \qquad (12)$$

Then the marginal cost schedule for a given level of output $Y=Y^*$ may be found by,

$$\frac{dC(Y^*,W,e)}{dY^*} \qquad (13)$$

The cost of the environmental input (for example, sedimentation) can be estimated by differentiating the marginal cost schedule with respect to e.

The relationships presented in equations (1) to (13) above can now be used to place values on the environmental effects of soil conservation projects. Despite the earlier discussion of the problem of nontraded goods and services, there are many ways in which market prices can be used to value these goods and services. And as will be shown many environmental changes have tangible effects on production (for example, changes in crop yield, fish catches) that are easily valued using market prices.

Appraisal Based on Changes in Net Income

This approach is used to examine situations where economic agents (that is farmers) operate in a setting characterized by varying environmental attributes and where all other factors do not vary or can be controlled. The analyst attempts to estimate the relationship between changes in farmers' net incomes and changes in environmental attributes.

This approach is often used where, for example, changes in yield resulting from changes in the rate of soil erosion (the environmental attribute) are used to estimate changes in net farm revenue "with" and "without" the soil conservation project. Numerous assumptions have to be made concerning acreage, cropping patterns and output demands.(eqn 1-4)

Appraisal Based on Changes in Total Cost

The basic objective here is to estimate the relationship between changes in environmental attributes and changes in producer costs.

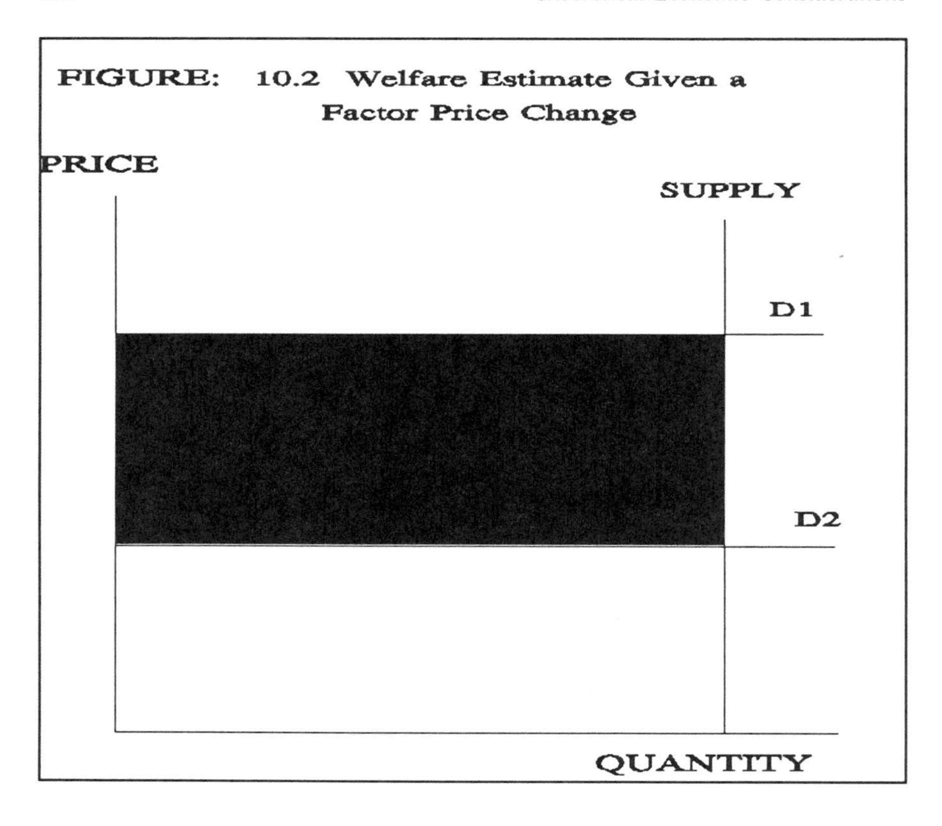

This approach requires data on costs, levels of environmental attributes and levels of other factors influencing costs. Such an approach is rather common as a way of examining the so-called "mitigation" costs of cleaning up drinking water impaired by erosion or where costs are incurred to replace the nutrient value of the eroded soils -- for example, replacing the nitrogen lost through erosion with of fertilizers. There are variants of this approach (equation (8)).

Appraisal Based on Changes in Input Value

The objective here is to estimate how the price of any particular input is affected by changes in soil erosion. The benefits or costs of a project designed to protect the soil are measured through factor markets -- for example, estimating the effect on land values of differences in topsoil depth or quality, this approach is sometimes called "hedonic".

This approach has been applied to changes in land values or in transportation costs (that is riverine transport), but it could also be applied to changes in other inputs. The analyst needs data on input prices along with data on environmental quality, the prices or other factors and factor or usage factor (equations (9) to (11)). Figure 10.2 illustrates a welfare estimate when this approach is used. There are many variants of this approach, some of which cannot be used in LDCs because of lack of data.

Appraisal Based on Changes in Output Supply

The objective here is to estimate the relationship between changes in soil erosion rates and shifts in the output supply curve induced by these changes. These shifts occur through changes in the marginal cost of production "with" and "without" the project.

This approach has not been used extensively, but it is useful for estimating changes in the per-unit cost of supplying hydroelectric power due to change in the rate of sedimentation in rivers or behind dams. It can also be used in cases where there is an increase in cost per unit of water sold when water pollution increases (equations (12) and (13)).

Getting Back to Welfare Estimates

The measures described above are not necessarily welfare measures. Several conditions have to be met. Here, we relate the measures to the notion of producers' or consumers' surplus. The major assumption in developing welfare estimates in this way is that the markets to be examined are in equilibrium "with" or "without" a project.

Estimates of a Change in Total Net Income

An estimate of net income must be related to a measure of producers' surplus, or economic rent. One will therefore assume that a change in environmental quality which results in a change in farm revenue leads to a change in producers' "rent", which is a true measure of welfare. However, several assumptions must be made for this to be an adequate measure of welfare. The principal assumption is that there is a totally elastic demand curve. With such a curve, shifts in the supply curve reflecting changes in net income become a true measure of producers' welfare. Thus, this kind of estimate can only be used when the analyst assumes that the project does not influence demand prices (the so-called "small country assumption").

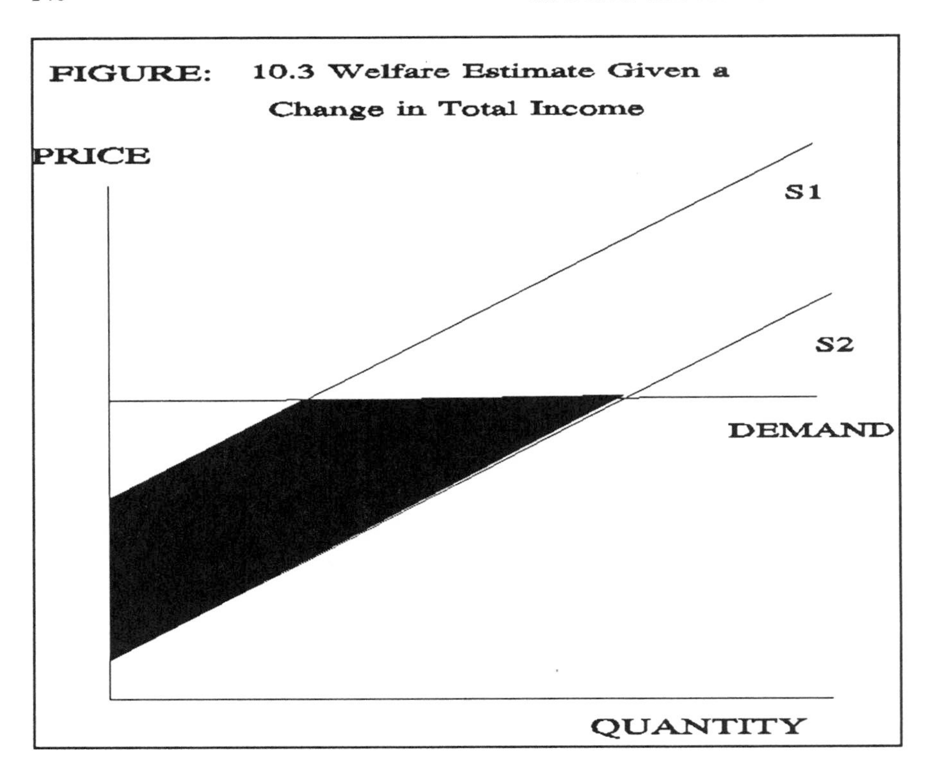

This situation is illustrated in Figure 10.3. There, the supply curves without and with the soil conservation project are assumed to be S and S‘, respectively. The total change in welfare due to the project is equivalent to the area between the SS and S‘S‘ curves below the demand curve DD (abcd). This change can be seen as a change in producers' surplus. Since consumer surplus is zero with a totally elastic demand curve, only the change in producers' surplus (net income) is relevant in developing a total benefit estimate.

Estimates of a Change in Total Cost

The change in total cost required to deliver a unit of output can be used as a measure of the welfare of producers and consumers. The main assumption here is that there is a totally inelastic demand for a commodity -- that is, a fixed quantity will be sold at any price. Therefore, when changes occur in the total cost of providing the fixed quantity, this cost can be equated to a change in consumers' plus producers' surplus.

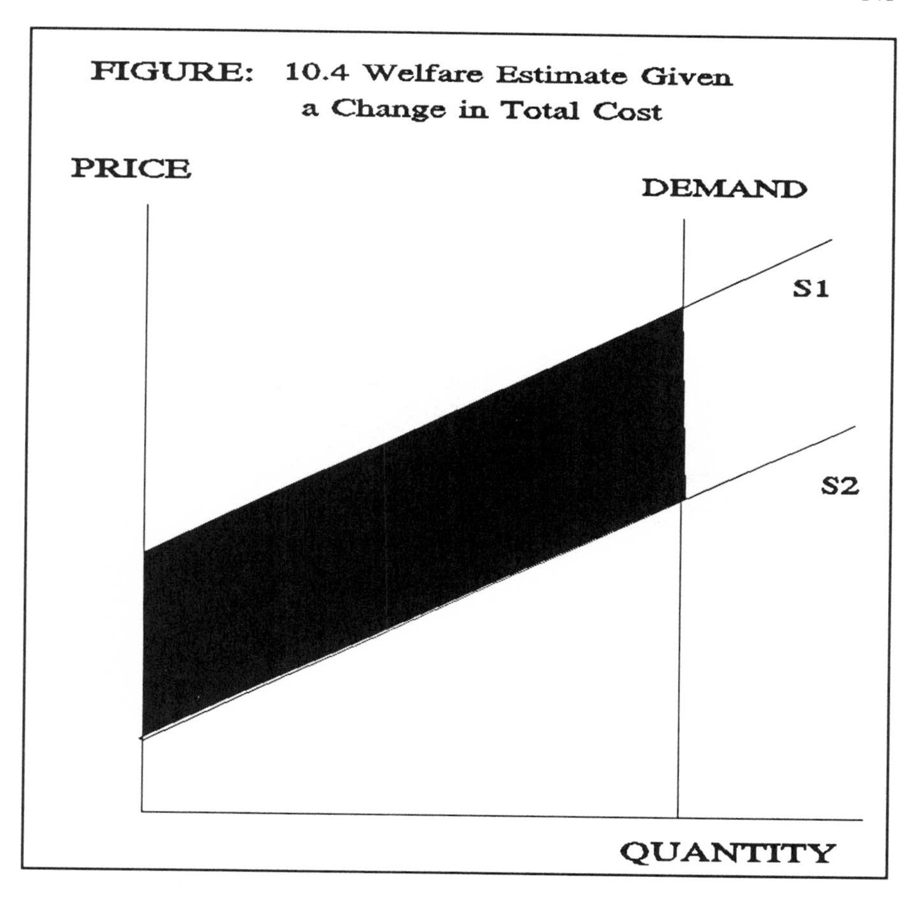

This situation is illustrated in Figure 10.4. The total change in welfare due to the project is shown by the shaded areas between the supply curves with and without the project.

A Point Estimate of a Change in Marginal Cost

When a farm is affected by a change in environmental quality, the analyst may also derive a welfare estimate using point estimates of the change in the price paid for any particular production factor. To obtain this welfare estimate, the analyst must often assume that the supply curve of the factor is totally elastic while the demand curve is totally inelastic. The change in producers' and consumers' surplus can be estimated by looking at the difference in price induced by a change in soil status multiplied by the quantity of factors consumed.

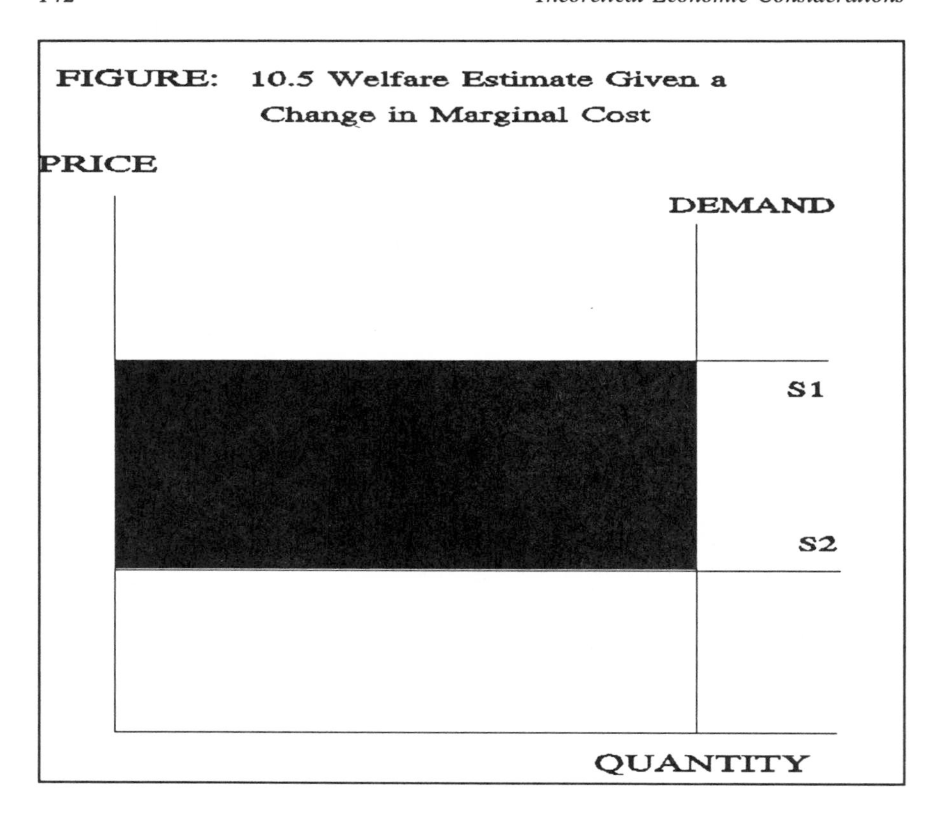

FIGURE: 10.5 Welfare Estimate Given a Change in Marginal Cost

This is illustrated in Figure 10.5. The resulting change in welfare is equivalent to the shaded area between the demand curve and above the supply curve, which shifts down as a result of the project.

Shift in the Supply of Output Schedule

The welfare effect of a shift in the output supply curve may be estimated by calculating the change where the supply curve intersects the demand curve. This involves the determination of changes in quantities and product prices. It is expected that both producers' rent and consumers'

surplus will change. This is illustrated in Figure 10.6. In this case, consumer welfare without the project is shown by area "a" and producers' welfare by "b" plus "e". Consumers' welfare with the project increases by [b + c + d], while producers' welfare is decreased by the amount of area b but increased by [f + g]. The net change for producers is [f + g - b]. The net increase in social welfare is [c + d + f + g].

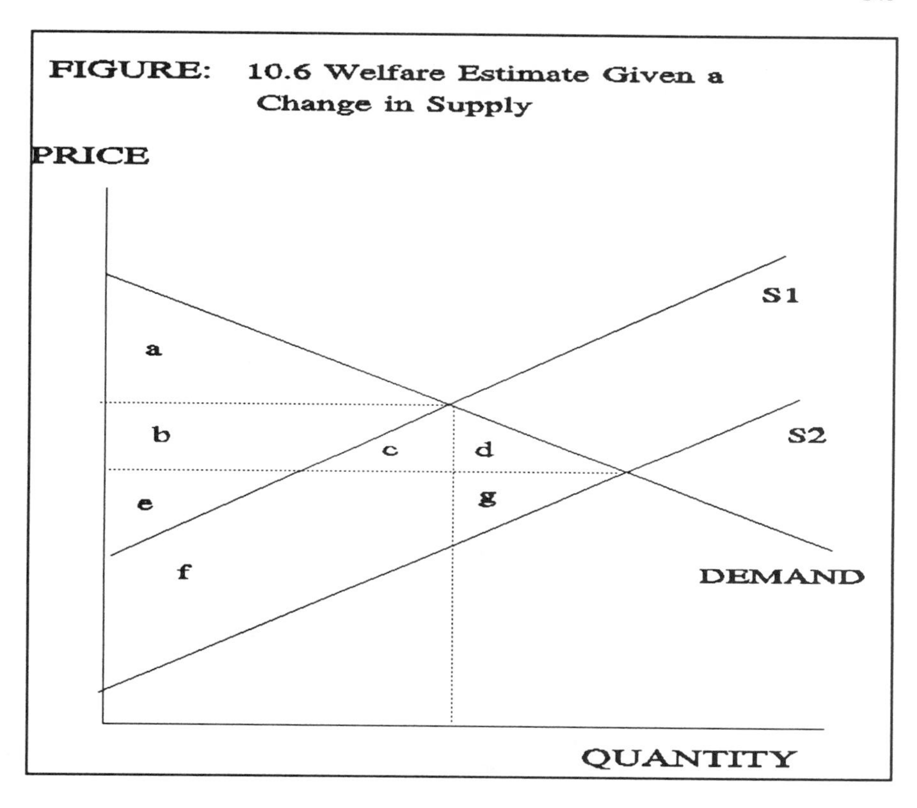

FIGURE: 10.6 Welfare Estimate Given a Change in Supply

Changes in the Factor Supply Schedule

Given an estimate of a change in factor demand, one can compute the change in social welfare by considering the effect on both producers and consumers, given the shift in the market equilibrium point. This is illustrated in Figure 10.7. The net effect of a project is a gain in total welfare [d + e]. However, this shows that factor owners gain [b + e], but that consumers gain d but lose b, so that net gain is [d - b].

Estimates of a Change in Welfare with a Change in Yield

With respect to soil conservation projects, one often attempts to estimate changes in welfare by multiplying the crop price by change in yield with and without the project. The major assumption here is that the demand curve is totally elastic and the supply curve is totally inelastic. This is illustrated in Figure 10.8. The shaded area is the change in total welfare.

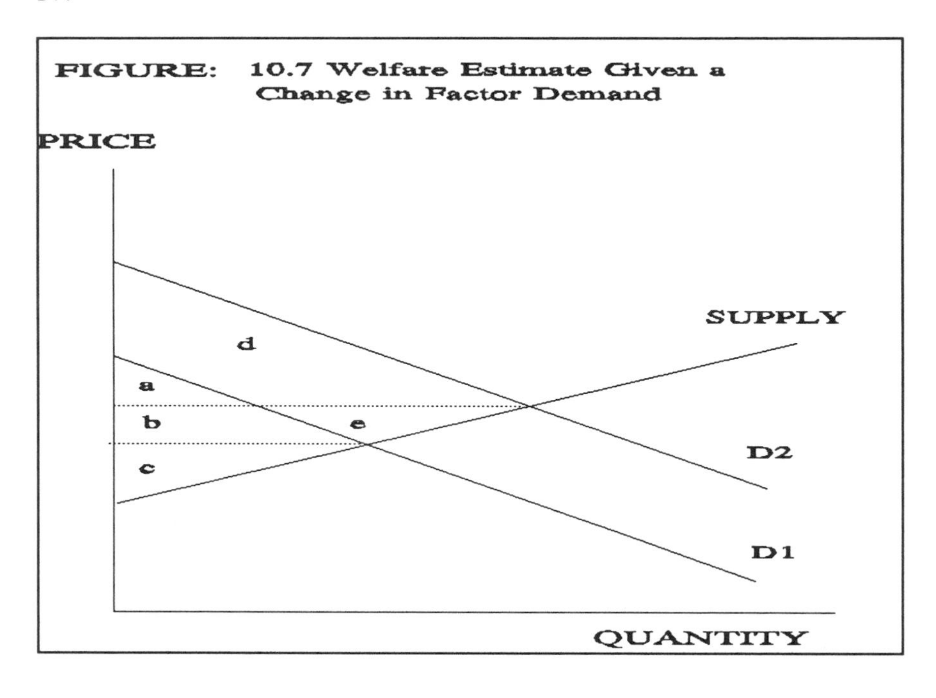

FIGURE: 10.7 Welfare Estimate Given a Change in Factor Demand

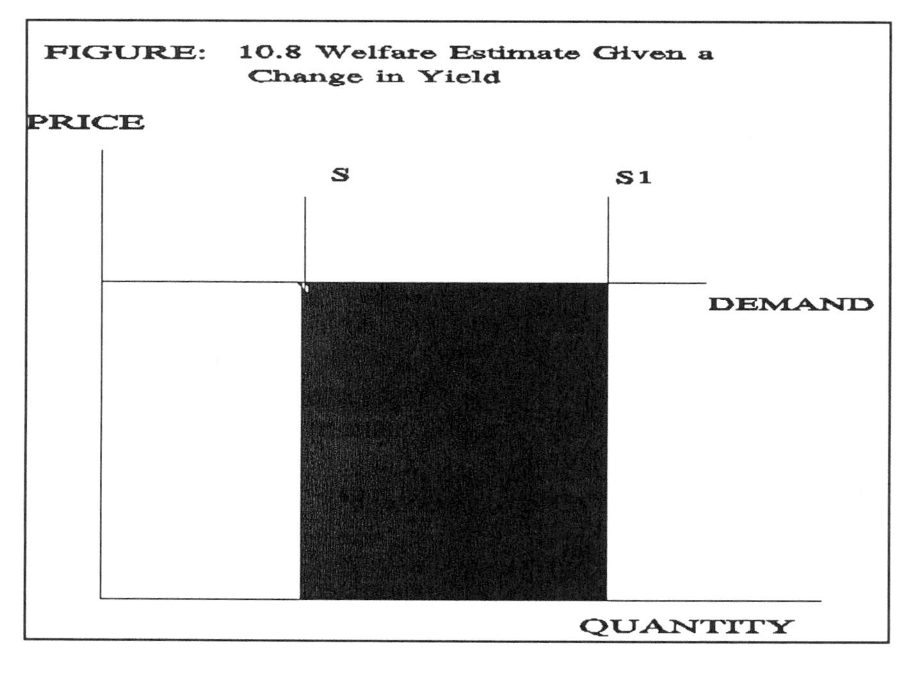

FIGURE: 10.8 Welfare Estimate Given a Change in Yield

Avoiding Double Counting

The welfare impacts of soil conservation projects are rather complex. These projects affect producers and consumers, directly or indirectly, through changes in input or output markets. Therefore, one needs to avoid double counting when applying different measures of welfare or adding up the project's main effects. This is particularly important when the analyst computes effects on the basis of net income, cost of a production factor and effect on output market.

Indirect or Secondary Effects

Soil conservation projects have several types of indirect or secondary effects. These are economic effects on other agents in the economy which are located outside the geographical boundaries of the project (for example, traders, other regions, urban centers). These effects may result from changes in the input market (for example, an increase in fertilizer demand may increase the efficiency of the fertilizer industry) or changes in the output market (for example, an increase in the production of any particular commodity may expand interregional trade). The magnitude of these effects depends, among other things, on the size of the project.

There is great controversy about the true extent of these affects or under what conditions they exist at all. Thus, the extent to which these effects should be taken into account in the economic analysis of any particular project remains in doubt.

When production factors are fully employed and there is no excess capacity in industries supplying the project and there are no potential gains (or losses) from economies (or diseconomies) of scale, secondary effects in one region are often offset by secondary effects in another region. From the viewpoint of the national economy, one could generalize by saying that these effects increase the welfare of society in "net terms" only if there are surpluses in a fixed or variable factor of production which will be utilized by the soil conservation project, or if an unutilized means of maximizing existing resources exists somewhere in the economy.

Secondary effects play a role in analysis of *distributional* effects. Empirical measurements of the magnitude are complex, however.

Valuation Methods

Soil conservation projects, like environmental projects in general, are difficult to evaluate because of the existence of unpriced attributes --

attributes not directly traded in markets. The preceding discussion indicated the general theoretical approaches to valuation, based on producers' and consumers' surplus. We now turn to various empirical approaches to valuation (Freeman (1979), Sinden and Worrell (1979), and Hufschmidt, *et al.* (1983)).

There are a number of methods for deriving the value of alterations in the environment due to soil erosion. These methods fall into three classes, (a) development of value estimates from observed economic behaviour, (b) development of value estimates from elicited responses (surveys), and (c) development of value estimates from synthesized or simulated economic behaviour.

Valuation Based on Observed Economic Behaviour

It is now common to develop welfare estimates by observing the economic behaviour of different parties as they are affected by changes in environmental attributes. This approach involves observing a number of individual producers over time, obtaining data on variables of interest (for example, prices, profits, costs) and obtaining data on changes in environmental quality (for example, erosion rates). Using these data, an estimate is made of the consequences of changes in environmental variables.

There are many ways to classify valuation techniques. One typology divides observed economic behaviour approaches (market-oriented approaches) into two groups -- those based on benefit valuation and those based on cost valuation (Hufschmidt, *et al*, 1983).

One type of *benefit valuation techniques* is based on the actual market prices of goods and services. The most commonly used approach in evaluating soil conservation projects is based on *changes in value of output* -- do crop yields go up or down? If soil erosion decreases and crop yield increases, a "with" and "without" analysis could define the extent of this benefit. A second type of benefit valuation technique is based on the use of *surrogate markets* to value an environmental change.

Table 10.1 Tableau of Benefits and Costs

	Benefits	Costs
Primary	Project benefits, derived directly	Direct project costs, Environmental costs
Secondary	Multiplier or benefits induced	Multiplier costs or costs induced

A subset of these are the hedonic property-value approaches discussed later under factor demand. Surrogate market approaches can also be very useful. The important thing to remember is that the analyst should use all information available, since changes can sometimes be valued from the benefit side and sometimes from the cost (or cost-avoided) side.

Many "cost approaches" are outlined in the literature. This type of approach is commonly used to study potential or actual changes in maintenance, mitigation, damage replacement, or damage prevention costs.

Soil erosion may affect road maintenance costs, irrigation investment costs, or the costs of hydroelectric power equipment. These effects are the results of such things as landslides, increases in water turbidity, or increases in the chemical content of water.

Soil erosion may affect the mitigation cost of a particular activity -- for example, drinking water supply. Increases in water purification costs arising from increases in sediment load or chemical pollution, or increases in sediment removal costs are examples.

Soil erosion may affect the timing of the replacement of assets or the replacement costs. When dealing with replacement cost, one should distinguish two types, (1) the replacement of services with services from a more expensive source and (2) the replacement of existing capacity. The *replacement cost approach* is often used when siltation of a water reservoir makes it necessary to replace hydroelectric power facilities with thermal sources of power. The second approach is used when a new dam, a new bridge, or a new road needs to be built because of damage caused by erosion.

The costs of downstream damages caused by soil erosion are usually reflected in changes in economic activity -- for example, in irrigation infrastructure construction, groundwater development, or activities like agriculture and fisheries. Increased rates of siltation in reservoirs reduce their economic life. Thus, soil conservation programs designed to reduce siltation rates expand the useful life of reservoirs. If no program is adopted, reservoirs will have to be replaced sooner than expected. Under the welfare conditions explained earlier, the cost of replacement can be used in estimating the potential benefits (that is "cost saving benefits") of the program. By the same token, replacement of tube wells, or differential pumping costs, can be accounted for as benefits of soil conservation projects, since they prevent further deterioration of existing aquifers.

Finally, soil erosion affects the cost of preventing damage due to erosion. The construction of sediment ponds to prevent sediment from entering river water, or the lining of stream banks with concrete to prevent floods, are good examples of prevention costs.

These cost approaches are based on the assumption that the same level of demand will exist, with or without a change in the rate of erosion because of the project.

The approaches should be used to value only the marginal change in environmental quality caused by the project.

With regard to *factor demand*, or changes in the marginal values of factors of production, observed economic behaviour can also be used to generate estimates reflecting environmental changes. This is commonly done. Within this hedonic approach land values are estimated as a function of a number of parameters, including the environmental quality resulting from different levels of soil erosion. One then isolates the impacts of alternative environmental quality levels and the corresponding price effects.

One may need to distinguish between (1) the market value approach (true "hedonic") and (2) the estimated value, or income foregone, approach in valuing property. Both approaches assume that the land market is in equilibrium and that land rent or land value reflects future income (that is productivity) from the land. They also assume that changes in the quality of land are visible and that the price of land will therefore change to reflect changes in land quality. These assumptions are sometimes unrealistic. Erosion may be difficult to see. Further, there is strong segmentation of land markets in rural areas. If land markets are "perfect", market and estimated values will be approximately the same.

Markets for land in LDCs are often inactive or very imperfect particularly in rural areas. Rural land values often reflect elements other than changes in soil productivity. Land is often purchased for security reasons, for land speculation, or to increase one's wealth or prestige.

In the absence of a market value, many soil conservation projects rely on an estimate of the implicit value of land that is obtained by determining changes in productive capacity "with" and "without" the project. In other words, by estimating the value of land as the discounted value of future income streams from the land. This is a "flow" value and is a measure that will tend to underestimate the true value of land to society. In countries where land is a major constraint to development and soil losses are such that irreversible damages are occurring, the reduction in the size of the country's stock of land, will put a high premium to each remaining unit of land. This is to say that irreversibility will also affect the stock value of land.

Changes in factor markets that are the result of changes in soil quality may also be estimated by examining changes in demand and supply schedules as directly calculated by profit or cost function estimation.

Valuation Based on Elicited Responses

This approach to valuation is based on data directly elicited from different factors in the economy. This nonmarket data is obtained through surveys, questionnaires, bidding games and voting. Since these techniques are not considered to be very applicable in LDCs, they are only discussed briefly here. This approach involves gathering data on such matters as individual willingness to pay for better water (by reducing silt or the concentration of chemicals) or for retaining the future productive capacity of land by reducing topsoil loss. Such elicited data, is directed toward the development of benefit estimates for changes in environmental quality attributes.

There are a number of conceptual and practical difficulties in developing demand equations from elicited data.

Valuation Based on Synthesized or Simulated Economic Behaviour

It is difficult to isolate each environmental factor that pertains to soil conservation while holding other factors constant. Consequently, a synthesized economic model is commonly used during project appraisal. The complexity of the model may vary from such techniques as linear programming, simulation and optimal control, to using farm budgeting to develop estimates of changes in profits.

There are many decisions and assumptions to be made regarding the structure of the model and the model's solution process. In a budgeting exercise, for example, one may assume that acreage and cost remain unchanged.

Assumptions and Development Dynamics

This chapter has discussed the theoretical basis of welfare measures and how it can be translated into specific appraisal frameworks. Several approaches were presented according to the method the analyst uses for gathering data. In analyzing these methods, emphasis was given to the major assumptions underlying them. Nothing was said however, about certain assumptions relating to the dynamics of the soil conservation system. Two of these are briefly mentioned here, technology adoption and discounting.

One major assumption under which most appraisals are carried out is that farmers will in fact adopt the techniques necessary to carry out a soil conservation project. If farmers are to adopt any soil conservation program, they must somehow be convinced that the program will lead to net improvement in their position. Decisions about adopting a program are made by farmers in environments characterized by profit maximization, risk avoidance and pressure to assure subsistence production. Due to the nature of soil conservation programs, most benefits accrue far in the future. Thus, subsidies or other compensation schemes will be necessary in many instances to assure adoption of new technologies.

Most of the approaches to be used fall in the category of benefit-cost analysis, referred to as discounting analysis. This issue is related to issues of intergovernmental equity, irreversibilities and uncertainties which are discussed in latter chapters.

Notes

1. Another way to present this is as follows. Commodities like soil loss and sedimentation are not traded in the market. Thus, no value is readily available to quantify benefits and costs. To avoid this problem, three steps should be taken: (1) conceptualize these commodities in terms of their characteristics (e.g. in the case of sediment these characteristics are volume, size, shape, and chemical content); (2) establish a clear relationship between changes in these characteristics and changes in productivity of tradeable commodities (e.g. via net income, profits, total costs, marginal costs, import costs); and (3) assess these changes in economic terms, using the prices and quantities of the tradeable commodities affected.

2. This section is based on Binswanger (1975).

11

Illustrations

Introduction

This chapter presents a series of case studies, or illustrations. The main objective is to show how technical and economic frameworks have been applied in the past. The presentation suffers from several limitations, due mainly to two factors. First, it is difficult to determine the appropriate analytical depth of each presentation. Illustrations showing the application of a technical concept are likely to appear simplistic to a natural scientist, but may be helpful to an economist or planning officer. Similarly, illustrations showing the application of economic concepts may be simplistic to the economist but useful to the natural scientist.

Second, the material from which these case studies were drawn were seldom written in a way that would make it possible to show many procedures in great detail. Those who are interested in more detail should consult the references cited in each instance.

The illustrations are clustered by subject matter. Thus, in one case study the reader will find several applications or the use of different approaches to a similar problem. Other case studies simply illustrate how a relationship was defined or how a particular method was applied. The case studies and illustrations are divided into two classes, upstream effects and downstream effects. In addition, some case studies deal more with the technical aspects of a method, while others concentrate on questions of economic valuation. In most cases the illustrations are drawn from material published elsewhere, which is generally available.

Illustration 1 is not a case study *per se*. Rather, it illustrates the use of economic analysis in the project design process. The next seven examples illustrate analysis of farm-level, on-site effects of soil erosion and its consequences. Some focus on technical aspects (TA), such as the soil-yield relationship, others focus on economic aspects (EA), such as placing values on lost nutrients:

Illustration 2:	The Economics of Soil Conservation in a Semi Arid Country (EA)
Illustration 3:	The Physical and Economic Impacts of Sedimentation on Fishing in a Reservoir in NE Thailand (TA & EA)
Illustration 4:	Yields and Depth of Topsoil (TA & EA)
Illustration 5:	Estimating Soil Life-Span (TA)
Illustration 6:	Economic Value of Nutrients (EA)
Illustration 7:	Replacement Value Approach (EA)
Illustration 8:	"The Paradox of the Plow" (TA)

The next two illustrations deal with soil management techniques and soil conservation projects at the on-farm level:

Illustration 9:	Costs and Benefits of Terraces (EA)
Illustration 10:	Varanasi Soil Conservation Project (EA)

The next two illustrations deal with the physical interaction between forests and water in the upper watershed, especially the effects of forest land conversion to other uses. These two cases are as follows:

Illustration 11:	The Influence of Forest on Water Runoff (TA)
Illustration 12:	Effects of Soil Conservation in India (TA)

The last seven illustrations are concerned with analysis of the downstream effects of soil erosion-sedimentation, water quality changes and flooding. As described in Chapters 5, 9 and 12, these effects are frequently valued as "costs avoided" if upstream soil erosion is controlled. The seven examples are the following:

Illustration 13:	Economic Aspects of Sedimentation (EA & TA)
Illustration 14:	Reservoir Sedimentation (EA)
Illustration 15:	BCA of Hydrometeorlogical Data (TA & EA)
Illustration 16:	Effect of Land Use on Flood Flow Peaks (TA)
Illustration 17:	Estimation of Flood Control Benefits (EA & TA)
Illustration 18:	Change in Water Quality (EA)

Overview

Illustration 1. The Use of Economic Analysis In the Design of Projects: A Land Clearing Project

The design of watershed management schemes involves choosing among conservation techniques and possible land uses. Generally, economic analysis is brought to bear on these projects after the range of alternatives has been narrowed to the point where only an "accept or reject" type of conclusion is possible. Although it is not feasible, or necessarily desirable, to attempt to design projects by selecting from an unlimited range of options, experience suggests that the role of economic analysis could be broader than it usually is.

The typical project brought to the financial or economic analyst can be described as the "limited case." Decisions have already been reached on the size, location and nature of most, if not all, components of the project. The end use of the land has been selected, a manual, mechanical, or biological technique has been chosen and the temporal and spatial scale of the project has been determined. Given this context, economic analysis basically involves appraisal of the overall benefits of the project.

Compare this approach with the so-called "expanded case." Here, selection of project components is regarded as the *result* of the analysis, rather than as a *constraint*. In the case of a land-clearing project, economic analysis can make greater contribution by providing a framework for systematically screening combinations of land-clearing methods, subsequent uses and projections of future consequences. The consequences of employing the "limited case" include:

(1) The likelihood that selecting project components on the basis of conventional wisdom may not hold up to systematic investigation.
(2) Neglect of off-site or downstream effects that are important in assessing natural resource-based projects.
(3) Failure to fully consider intertemporal concerns, such as intergenerational equity and distribution. And
(4) Difficulty in relating individual projects to overall development objectives.

For example, land-clearing schemes in West Africa designed and appraised according to the "limited case" model have typically been based on the following seemingly plausible assumptions:

1. That labor is in short supply and therefore both manual land clearing and manual farming operations are too costly to contemplate on any large scale.
2. That sustained annual food cropping must be developed rapidly and on a large scale from existing forest lands.
3. That the only technology which meets both of these conditions is mechanized land clearing, followed by mechanized farming.

Implicitly, it is also assumed that the project will generate insignificant externalities, will be of relevance only to the current generation and will be isolated from any larger process of expansion of the land frontier.

Consideration of the basic assumptions above illustrates the potential role of an "expanded case" model in the design of a project. For example, with regard to the presumed labor constraint, one might ask: What is the supply function of labor? Can labor be trained to be more productive? Are there profitable techniques of manual farming? Can management be trained to handle large labor gangs more efficiently? The answers to these questions may significantly expand the range of potential techniques. In the case of upland watershed management, a similar analysis might broaden the scope of available choices from fixed engineering structures such as terraces, to biological and nonstructural erosion control measures, such as mulching, contour ploughing, or feeding of cattle in their stalls.

Similarly, the assumption of sustained annual cropping as the ultimate use of the land can be questioned. This will affect not just net proceeds but may also lead to the relaxation of constraints on choice of land-clearing technique. For example, choosing between tree crops, pasture, or periodic cultivation with extended fallow periods would allow consideration of various land clearing techniques, such as manual ones that remove less of the existing cover. In addition, consideration of alternatives to mechanical farming allows specific consideration of smallholder schemes that might distribute benefits in more desirable ways.

Moreover, a broadened approach to planning would allow consideration of downstream effects that are typically not quantified in the limited case. Elements that would broaden the economic analysis would include patterns of soil loss under alternative clearing and management systems, potential downstream consequences of changes in sediment loads and flow regimes which might follow large-scale land clearing in upper catchments, the probability that certain management options, although viable on the basis of, say, a 20-year project life, might lead to irreversible situations and the foreclosure of options for future generations, and inter- and intra- generational conflicts over income distribution stemming from differences between those benefited and those disadvantaged by alternative plans.

One might conclude from the foregoing that economic analysis of land-clearing methods and subsequent agricultural or forest management requires physical and social scientists to identify a large array of relationships and to predict how interaction between natural and economic systems would affect production and income distribution. But everything is not related to everything else. Simple simulation modelling can be useful to decision makers in assessing key interrelationship which influence the economic outcome of alternative approaches.

The economist's contribution to decisions on soil conservation methods will vary, depending on how many elements in the decision framework have been fixed *a priori* and how many are available. A flexible approach necessitates a search for relevant alternatives. Constraints are determined by overall development goals (for example, increased per capita consumption or income distribution) and not by prior decisions on means (for example, specific crops or production methods). It is axiomatic that as the number of variable elements increases, the analysis becomes more complex and requires greater interaction between the economist, other social scientists and physical scientists.

Over the past 10 to 15 years a great deal has been learned about the behaviour of physical systems as a result of different land-clearing methods and post-clearing management techniques. Indicators of relevance to economic analysis include, yield behaviour, fertilizer response, response to cultivation versus minimum tillage, fertility, recovery under fallow, soil loss and irreversibility of soil degradation. For purposes of economic analysis, data on the key input-output relationships of these technologies should be systematically collected for different situations over extended periods of time. Similarly, it is important to encourage research on institutional modifications to fit appropriate resource management systems.

In order to develop guidelines for further research, one can draw on a considerable body of knowledge. For example, there is now documentation that explains the relationship between clearing methods plus post-clearing management systems and soil erosion, flow regimes, compaction, fertility, sustainability of yields and the costs of maintaining yields and sustaining permanent agriculture. The physical environment which conditions these relationships is now reasonably well understood. On the basis of different case studies and trials, informed judgements can now be made on the institutional factors that influence tropical land development schemes -- for example, factor costs, labor productivity, management capacity, information flows, land tenure, price instability, input supply and marketing.

The issues relevant to sustainable resource management can only be formulated and analyzed within the dynamic context of the "expanded case." That is, the implicit assumptions of the typical limited case should be made explicit and then examined carefully. Economic analysis can be used to help formulate and assess a wider range of resource management options and provide indications of cost-effectiveness (economic return) as a basis for identifying critical physical or institutional factors which influence economic performance. Future research should be directed toward determining these factors.

On-Farm Effects

Illustration 2. The Economics of Soil Conservation in a Semi-Arid Country[1]

This study examined the benefits of soil conservation activities in Mexico. Even though six types of benefits were identified (increased productivity, production cost reductions, sustained soil productivity, reduced off-site losses, reduced out-migration and job creation), it was found that no hard numbers on external or off-site benefits were needed because the on-site benefits were large enough by themselves to justify soil conservation programs. Any off-site benefits from soil conservation programs only improve the economic viability of the programs.

The study focused on the protection of rainfed agricultural lands producing corn and beans. The soil conservation practices examined included terracing, contour ridging, gulley check dam construction and hedge planting. The costs per hectare for programs of terracing and check dams in 1976 ranged from US176 to US$211 for work undertaken by manual labor and US$111 to US$139 for similar work using tractors. Hedge planting of Nopal or Maguey along ridges increased costs from US$64 to US$90 per hectare.

These programs were designed to sustain agricultural production by controlling erosion. It was assumed that, without these programs, increasing rates of soil erosion would have required the abandonment of the land in 15 to 25 years. An internal rate of return calculation was carried out for two areas in Mexico, the IRR ranged from 9-15 per cent to 11-18 per cent, depending on how many years it might take before total soil loss would occur without the program.

As explained by Schramm, "total benefits over time consist of two components. The first is the addition to net output of the protected land minus the maintenance and repair expenses for the soil control measures.

The second consists of the prevention of soil losses and hence, the reduction in the value of the original net output (that is, value of gross output minus farm production costs). In the absence of more accurate information it was assumed that this reduction in basic output would proceed linearly up to the point of total soil loss or land abandonment."

This analysis showed that in this case soil conservation measures paid for themselves, even if only on-site effects were included in the analysis. The on-site effects were related to sustaining existing yields by preventing continuing soil loss. If off-site effects were added in, the economic return (and the IRR) would have been even higher. In addition, there were employment and income distribution benefits from the program.

Why didn't farmers adopt these measures voluntarily? Schramm says that "the lands and soils that have to be protected are almost completely in the hands of subsistence type farmers. They have neither the resources nor the know-how to invest in the protection of their land. Moreover, the largest percentage of benefits obtained consist of the maintenance of the productive capacity of the land, increases in net output are less significant. Hence there is no real source of new income to the farmer himself that could be used to amortize the original soil protection investment costs."

This was a clear case where an economically justifiable program required government assistance to become operational. The social benefits clearly exceeded the social costs. Although the economic and technical aspects are clear and known, there have been major implementation problems due to budgeting procedures and institutional constraints.

Illustration 3. The Physical and Economic Impacts of Sedimentation on Fishing Activities in a Reservoir in Northeast Thailand[2]

The Nam Pong Watershed in northeast Thailand provides water to the Ubolratna Dam, a multipurpose project designed to produce electric power, irrigation water and flood control benefits. Since the dam was completed in 1965, there has been extensive forest destruction in the watershed of the Nam Pong Basin.

Erosion resulting from illegal forest destruction has increased sediment discharge into the reservoir. This inflow of suspended sediment has resulted in increased turbidity and sediment deposits within the reservoir. These, in turn, have had negative effects on the survival of fish fry and fish growth. The study reported here focused on the physical and economic impacts of sediment on an established fishing industry located within the reservoir, other economic effects are not included.

An integrated simulation model of the Nam Pong Basin was created. The model consisted of four interacting sub-models covering the dynamics of water management, land-use patterns, socioeconomic factors and reservoir fisheries. Sediment outflow from the basin was derived from a detailed watershed sub-model. As the sediment loads of each stream were passed on to the reservoir, they were changed to turbidity levels. These levels, in turn, were used in the fishery sub-model to determine the impact to turbidity on fish fry. This sub-model relates nutrient inflow, plankton growth and fish production. Turbidity had a negative effect on survival.

The Nam Pong Reservoir fishery yielded about 1850 tonnes of fish per year during the 1965 to 1975 period. As soil erosion in the watershed increased however, the average total yield began to decline even though the number of fishermen increased. Thus, fishery return per unit of effort was significantly reduced.

This study compares actual and predicted fish harvests and economic returns for selected years. It estimates that economic returns per fisherman in 1990 will be only about 60 per cent of the 1969 returns if current trends continue.

Several management alternatives are considered. Reforestation and watershed management activities would help control the amount of sediment flowing into the reservoir. This inflow has already increased over threefold from the rate of 950,000 tonnes per year when the dam was first closed. If sediment inflow stabilized at three million tonnes per year, it might be possible to avoid a complete collapse of the fishery if overfishing could be controlled. This would require fishery management as well as limiting the number of fishermen to around 2500, a decrease from the present level. At that level, annual gross revenue from fishing would be around $500,000, or $200 per fisherman.

Although it was not done in this study, it would be possible to estimate the annual value of the fishery industry in two scenarios, with and without a watershed management program. This information could then be used to determine the difference in economic return between the two scenarios. Once this difference was analysed, the value could be used to justify watershed management expenditures.

A complete restoration of this facility is unlikely because of institutional factors affecting both watershed and fishery management. This case illustrates how data on a marketed resource (fish) can be used to estimate part of the "cost" of reservoir sedimentation. This information can then be used to assess the benefits and costs of soil conservation and watershed management activities. The fact that the reservoir fisher is, in effect, an open access public good complicates the analysis.

Illustration 4. Yields and Depth of Topsoils[3]

Unless we know with some degree of certainty the relationship between erosion and productivity, appraisal of on-farm effects becomes an impossible task. There have been several attempts to analyze this relationship. Some have tried to estimate this relationship directly by correlating losses in top soil and yields. This method suffers from many limitations, since, as we already know, many other factors (for example, technology, fertilizer use and the like) influence yields and mask the effects of losses of topsoil on crop yields.

The general relationship is described as follows, the shallower the topsoil the lower the yields of a given crop. Although technological progress has increased yields in most soils regardless of depth, the increases have been greater in deeper soils. If one looks at this problem in a longer time dimension, it is possible to find that, despite technological progress, yields may actually decline.

An Empirical Assessment.[4] An economic analysis of the merits of soil conservation was carried out in Iowa, USA, where the principal explanatory variable to explain changes in yields was again depth of top soil. In the case of maize, the following relation was found, in a mathematical form:

$$Y = 1{,}670.49 \text{ x } X^{0.3} \qquad (R^2 = 0.99)$$

where, Y is yield (in kg) and X is topsoil (in cm).

The data comes from sampling fields with identical rainfall, soil erodibility, slope length and slope gradient. The relationship came out to be almost linear. The underlying assumptions were as follows, (1) yields have slope of 3%, (2) each year erosion leads to the loss of 50 tonnes of top soil (0.34 cm of top soil per year), and (3) 0.0254 cm of top soil are generated per year. In 80 years yields decline from 4,514 to 2,270 kg/ha equivalent to a drop from $451 to $227 per ha per year. The function expressing changes in yields over time is,

$$Y_m = 1{,}670\ (sd_n - t_n \text{ x } slc)^{0.3}$$

Where, sd_n = soil depth in year 0 (that is, 30 cm), t_n = year n, slc = constant net soil loss each year (that is, 0.34 - 0.0254 = 0.3146 cm/yr)

Thus, the yield loss until year n is then obtained through the following equation,

$$Y_{1m} = \Sigma^{n}_{i=1} 4{,}634 - 1{,}670 (30\text{-}t_m \times 0.3146).$$

A threshold value (NPV) of 470 was found. Therefore soil conservation practices that maintain the soil at the original depth (and yield level) and that cost less (in terms of NPV) are economic. If they cost more, it is economically more advantageous to the farmer to accept soil erosion and use up the soil resources.

A Damage Function Approach.[5] This approach consists of finding a "damage function" when the basic technical framework can express the relationship between crop yields and depth of topsoil. The analysis assumes that the farmer is using conventional tillage and is considering switching to a recommended conservation practice. The *erosion damage function* is a framework which enables analysts to compare the private profitability of changing from a conventional practice to a conservation practice. This function may be defined as:

$$D_t = P_e - P_c$$

Where D_t is the value of the damage function in year t, P_e is the private profitability of conventional practice, P_c is the private profitability of conservation practice.

This function was applied to an annual cropping area of the Palouse in Idaho and Washington. In the Palouse, (1) the major crop is winter wheat (in rotation with dry peas), (2) soils are loessial, (3) slopes as steep as 50% are cultivated, (4) average precipitation is 22 inches, (5) average annual loss in top soil is 14 tons/acre and reaches 100/tons/acre in certain areas, (6) soil surveys show that 10% of the area has lost all of its original topsoil during the last 100 years, and (7) 70% of annual precipitation occurs between October and March.

Under these conditions, minimum tillage was proposed as a conservation practice to reduce annual soil loss on most slopes to 2 and 3 tonnes per hectare. A damage cost function was used to simulate losses in income under different scenarios. The function that estimates changes in wheat yields (Y) with topsoil depth (X) was:

$$Y = 36.44 + 47.01 (1 - e^{-0.09864X})$$

The analysis shows that, in some soils, conventional tillage is more profitable than minimum tillage from a short-run perspective. However, it is unprofitable when long-run yield damage is considered.

Illustration 5. Estimating Soil Life-Span[6]

Soil science researchers have developed the concept of *"soil life-span"*: how long a soil will remain productive. This concept is not only important in terms of providing a more meaningful indicator than simple erosion rates and could be easily incorporated into development planning. It is also relevant in terms of defining the "*critical zone*" of soil resources.

Earlier, we criticized the fact that results are often presented in terms of losses of topsoil (for example, tonnes/ha/yr). From the viewpoint of economic analysis, one is interested instead in *productivity*. If the analysis focuses only on "on-farm effects," loss of topsoil is not a meaningful figure. What is relevant is the *productivity* of the soil that is left and how soil erosion rates affect that level of productivity. Topsoil losses are more relevant to the quantification of downstream effects (for example, sedimentation).

The concept of soil life-span focuses on an indirect *notion of productivity*. This case study describes a framework that was applied to estimate the life-span of several soils in Zimbabwe. Once the life-span has been estimated, policy makers can focus on how to expand the life of those soils.

Key assumptions in applying the framework are that, (1) rooting depth is used as the measure of the ability of the soil to sustain productive agriculture, and (2) once soil depth diminishes to minimum rooting depth, erosion is said to have rendered the soil nonproductive. Based on those assumptions, the following mathematical model was used,

$$L_f = \frac{(D_e - D_o)M}{Z - Z_f}$$

Where, L_f = soil life span in years, D_e = depth of available productive soils (m), D_o = minimum soil depth (or effective minimum rooting depth) for a particular crop (m), M = bulk mass of the soil (t/ha/m depth of soil) (i.e, bulk density in g/cm3 x 1000), Z = predicted rate of soil loss (t/ha/yr), and Z_f = estimated rate of soil formation (t/ha/yr).

The model was applied to the Sabi Valley in Zimbabwe. The value of M = 15,000, D^e = 0.3, Z^f = 1 and the level of minimum soil depth for a traditional grain crop like sorghum is approximately 0.15 meters and for a moisture-sensitive crop like maize it is 0.25 meters and the estimated rate of soil erosion was estimated at 80 t/ha/yr (using SLEMSA). The results were alarming. The number of years before the soil becomes too shallow to support maize is,

$$L_f = \frac{(0.3 - 0.25)\ 15{,}000}{80 - 1} = 9.5 \text{ years}$$

and for a subsistence crop of sorghum,

$$L_f = \frac{(0.3 - 0.15)\ 15{,}000}{80 - 1} = 28.5 \text{ years}$$

These soil life-span models give predictions which emphasize the serious threat that erosion poses to the agricultural potential of the Sabi Valley.

Illustration 6. The Economic Value of Nutrients

In two cases presented earlier, the method of monetary valuation of benefits and costs identified "depth" as one characteristic of soils -- and changes in soil depth were correlated with changes in yields (that is, a way to define productivity). These were changes in yields of tradeable commodities, to which the market assigns values. The illustration also pointed out that depth was a nonrenewable element of the soil.

In the material presented below, we focus on another characteristic of soils -- the availability of nutrients. In this case, the independent variable is a renewable element of the soil. The main objective of the presentation is to estimate the value of tradeable output foregone, or the farmers' income foregone due to loss of soil nutrients.

One of the best-known methods of estimating the value of income foregone has several steps: (1) estimate the amount of nutrients that are washed off due to erosion, (2) assess the relationship between nutrient losses and crop yields and (3) value the loss of income due to the decrease in productivity.

A general approximation for quantifying benefits foregone is the estimation of a "replacement value" of nutrients. The main objective here is to determine the total farm expenditure that would be required to replace the lost nutrients. In this particular case, the assumption is that the expenditure will be made to keep the level of productivity constant.

In a very few situations, one finds that an "hedonic approach" (that is, land value approach) is applied, as shown below. The assumptions stipulated in using land values to determine the benefits of flood control projects also apply in this case. One value of land is its productivity and a characteristic that effects productivity is the nutrient content of soils. This approach is presented below.

Land Values and Fertilizer.[7] In making decisions regarding land purchases, farmers use several criteria, among them debt-equity ratios and their ability to pay for the land, the level of productive capacity and the expected return from land investment. The use of fertilizers (that is, nutrients) affects farm profitability and in turn, will affect the price an individual farmer will be willing to pay for a given piece of land.

This case focuses on two types of soil in the Saskatchewan soil Zone (Western Canada), gray luvisol (most infertile) and thick black (most fertile). The methodology consists of determining the optimum levels of fertilizer use in these two soils and then comparing respective land values. The working hypothesis is that if fertilizer use has not been accounted for in the market price for land in the gray luvisol soil zone, investments in this zone will be preferable to investment in the thick black soil zone.

The land areas under analysis have the same rainfall patterns and therefore comparisons of crop yields are possible if fertilizer (particularly N and P) is used to overcome the nutrient deficit. Assuming no major market imperfection, or that any imperfection will affect land markets in the two areas equally, the market price of land will be determined by what farmers are willing to pay to obtain the economic returns from that land. This assumption allows the analyst to isolate the impact of different productivity levels on land price.

To determine the productivity of the soil type in each area, a base yield "without fertilizer" was established. To compute the difference in yearly returns from the two types of soil, the value of fertilizer was deducted from the value of expected yields (in different crops and different rotations). This assumes that all other costs of production are equal in the two areas for the same crop or rotation system.

At a 9% discount rate and using a planning period of 30 years, the difference in capitalized value between the gray luvisol and thick black soils was estimated to be $593 per hectare. Differences were also estimated for discount rates between 9% and 16% and for a 20-year, 30-year and infinite planning periods, respectively.

Land transactions were recorded and reduced to dollars per hectare. The difference in the highest transaction in each area was $525 per hectare, while the differences in the means was only $405 per hectare. The theoretical value may be higher or lower than the actual value, depending on the discount rate and the length of the planning period.

The major conclusion of the study was that where the highest market price are paid for land parcels in both areas, gray luvisol land is overpriced if the expected rate of return is less than 11.9%. For a finite planning period the break-even discount rate was slightly lowered.

Illustration 7. Losses in Productivity

Replacement Value Approach.[8] Icheon and Gochang, in the uplands of Korea, are the geographical areas covered by this case study. Land has been cleared to expand crop production areas because of population pressure. The principal cropping pattern in these areas was, barley to soybeans and sweet potatoes to barley -- but in the recently it has been changing to fruits and vegetables in order to reduce the rate of erosion.

The Universal Soil Loss Equation (USLE) was used to estimate the amount of topsoil lost in any particular year. With an erosivity factor of 500, an erodibility factor of 0.25, a slope length of 1.2 by 6 meters, a slope of 15%, a cropping factor of 0.35 (for barley/soybeans) and a farming practice factor of 0.76 (for contour terrace), the erosion rate was estimated at 40 tonnes/ha per year.

The replacement value, or the expenditure needed to maintain productivity at existing levels, was estimated by looking into the corresponding levels of nutrient losses. The annual nutrient losses per hectare were as follows: 15.7 kg of N, 3.6 kg of P_20_5, 14.6 kg of K_20, 10.6kg of Ca, 1.6 kg of Mg and 75.4 kg of organic matter.

Rice farmers downstream experienced siltation of their paddy fields from upland soil (108m^3 per hectare of upland development). In the application of the replacement value approach to assess different ways of restoring the damages, two strategies were considered. The first strategy was to asses the economic merits of bringing the eroded soil back to the uplands and with it the nutrients, by cleaning up the silted paddy fields downstream. The second strategy was to apply various soil management techniques in the upland areas (for example, straw mulching). Comparison of costs and benefits, using economic prices, favoured the second strategy.

In areas where the cropping system includes high value crops however, the first strategy may be more profitable. This applies to both the developed (for example, Holland) and the developing countries (for example, Indonesia).

Illustration 8. "The Paradox of the Plow"[9]

The rate at which soils erode depends, amongst other things, on the type of tillage. Although tillage is one of humankind's oldest forms of cultivation, the way in which it has been carried out has evolved from very simple handtools to wooden ploughs to iron and steel shares. Tillage is adopted most often for fine seedbed preparation, eliminating competition from weeds and improving the physical condition of soils.

Many problems, including erosion, are associated with different forms of tillage. Erosion effects are exacerbated when farmers plough up and down the hill. This has been called the "paradox of the plow". Conservation tillage is defined as contouring and planting across the slope (rather than up and down), to provide rows ridges and wheel tracks which act as barriers to the flow of water. One extreme of conservation tillage is zero or no tillage. This form of tillage reduces soil erosion by protecting the soil surface from the energy impact of raindrops, decreasing the velocity of runoff, improving water infiltration, reducing compaction and preserving soil moisture. Conservation tillage reduces not only labor (less ploughing) but also such considerations as fuel and energy use.

Conservation tillage also has disadvantages, the most frequent being the, delays in soil warm-up capacity, a slower onset of seed germination, higher rodent population, greater disease incidence and, interference with herbicide effectiveness.

In the USA, no-tillage area increased from 3 million acres in the early 1960^s to near 50 million in 1981. Because of the wide adoption of this practice, several studies have been conducted to study alternative forms of tillage, (1) molboard plough, (2) chisel plough, (3) disk plough and (4) no tillage. In one study, these different methods were applied to four adjoining fields for ten years and changes in soil properties were assessed. The study concluded that the field operated under no tillage had the most soil water, the least soil air, the greatest soil density, the least porosity and highest soil strength. Opposite results were found in the field where disk ploughing took place.

Illustration 9. Costs and Benefits of Terraces[10]

As explained earlier, soil conservation methods are of two types -- that is, biological or engineering [mechanical] methods. Because of topographical factors, the construction of terraces is often suggested, but policy makers do not always know if it is economically worthwhile for the farmer to do so. The object of this study was to assess the costs and benefits of terraces for erosion control -- in this case in Illinois.

Gross erosion from agricultural lands in Illinois exceeds 181 million tonnes annually and only 14% of sloping croplands are adequately protected from erosion. The study focused on the return to investments in terraces from the farmers point of view. The study involved several variables, range of soils [initially productivity, erodibility, kind of subsoils, slopes], soil loss calculations, estimates of yield reduction, terracing costs, and annual costs.

The USLE was used to estimate the amount of topsoil loss under various conditions. Such factors as rainfall, cropping management and erosion control practices were held constant. Soil loss was converted to volume using a bulk density of 84 pounds per cubic foot. Severe erosion was defined as a loss of 6 inches of top soil.

In calculating costs from the point of the farmer only, the analyst included a subsidy for terracing given by the government, which at the time was 50% of the cost up to a maximum of $2,500 per year. The study found that five factors had an effect on the cost of the terrace system over a 20 year period, kind of subsoil, slope, type of management, type of terrace used and soil erodibility.

The study concluded that if only the direct benefits to farmers were considered, most farmers would lose income by investing in terraces.

Illustration 10. Varanasi Soil Conservation Project[11]

Here we describe an *ex post* evaluation of conservation practices carried out in the Varanasi District, Uttar Pradesh, India. The soil conservation program in Varanasi started in 1964-65 and included contour bunding, levelling and check damming, as well as an extension program to advise farmers about the merits of these practices. At the time the evaluation was written in 1973, the program covered 20 units within the district, equivalent to an area of about 53,000 hectares. The evaluation concentrated on one unit, Betawan, which was thought to be representative of the whole district.

Three villages were selected at random at the start of the study and the soil conservation efforts of these villages were compared with a control group of villages that did not carry out the above mentioned practices but had similar ecological conditions. After a sample of farmers was selected, the major control variable was farm size. Farms were grouped into 2 hectare, 2 to 4 hectare and more than 4 hectare farms.

Only on-farm erosion effects were analyzed. In order to relate erosion to productivity, soil conservation practices were appraised in relation to a) increase in cropping intensity and b) replacement of inferior crops by superior crops. It was explicitly recognized that an increase in yields could not be attributed in totality to conservation practices. Use of fertilizers and improved seeds were also thought to be important contributors to productivity. The analysis was therefore not "pure" enough to determine the precise contribution of conservation practices. The evaluation technique was a farm budget type.

The major findings of the study were that a significant increase in intensity of cropping took place and that investment in inputs increased. The increase in cropping intensity was 57%, while the increase in input investments 97%. The net variance in the value of output was 112%. The resulting benefit-cost ratio was estimated at 4.6 to 1.

This approach is simple and illustrates how soil conservation programs can be evaluated. Because it is simple however, it has several limitations that should be noted. First, no downstream effects were studied. Second, the analysis was static and amortization of costs was assumed to take place in one year. This approach does not enable us to know the extent of diminishing returns on conservation practices. Third, because of the above mentioned attribution problem, the B/C ratio may have overestimated the return on conservation practices. And fourth, the B/C ratios for different sizes of farms varied. The study omitted an explanation of the net inputs in 4 hectare farms, thus giving the impression that it is less beneficial to introduce conservation practices the larger the size of the individual farm.

Illustration 11. The Influence of Forests on Water Runoff[12]

It is generally recognized that in basins characterized by high runoff, upstream forestation can be an effective method of flood prevention and control. The case described below however, demonstrates that economists need to consider the likely impacts of several characteristics of forests assessing their economic value.

Forests have been found to be impressive regulators of surface runoff, particularly during floods. In the typical forest-covered basin, runoff begins later and lower flows are recorded. On the other hand, runoff continues for a longer period of time in forested basins than in unforested ones. The volume of runoff generally is smaller as the size of the forest increases. But runoff also depends on a forest's types of trees and their ages, the season-related status of vegetation growth, the amount and intensity of rainfall, the soil characteristics and the amount of forest litter covering the soil surface.

Mita compared runoff in five sub-basins with roughly equal areas, altitudes and mean slopes. Forest coverage in the sub-basins varied by types of trees and by intensity, forest coefficients ranged from 0.0% to 95.4%. From observations of precipitation and runoff, three factors emerged as especially important in determining runoff, the degree of cover provided by the canopy (or interception), the amount of forest litter and the characteristics (that is, amount and intensity) of the rain. Interception is defined as the difference between the amount of precipitation recorded in open terrain and the amount reaching the ground in a forested area.

The proportion of precipitation retained by the forest canopy was higher the lower the amount of precipitation. Amounts of 1 to 1.5 mm per minute or less were entirely retained by the canopy. In heavy rains (of more than 100 mm), forest retention was only 3 to 4% of the amount falling on open terrain. The saturation of forest litter apparently is the major explanation for this decline in runoff regulation in heavy rains. This was found to be true for both coniferous and deciduous forests.

Finally, there appears to be a lag effect in the influence of both afforestation and deforestation on runoff. The remaining litter retards runoff in recently deforested areas, while runoff in recently of rested areas continues at high levels because litter has just begun to accumulate. The impacts of an ecological change in a forest, therefore, may not be immediately observed downstream.

Illustration 12. Effects of Soil Conservation In India[13]

The deforestation of upland areas for conventional agriculture road construction, mining and other uses has resulted in extensive and severe erosion in some countries. High rates of soil erosion have been reported in India, for example, even in watersheds classified as protected forests. People in these areas are permitted to exercise their grazing and lopping rights without adequate enforcement of provisions for the protection of soil. In addition, there is insufficient awareness among the people about the importance of maintaining grass and vegetal cover on the land surface.

Damage to life and property in valleys vulnerable to floods can be extensive. This illustration reports the results of an investigation of two conservation measures (bunding and check dams) as methods of controlling water runoff and soil loss.

Three watersheds were select for study, all of them in the Doon Valley of India, in the Himalayan foothills. The control watershed (W3B) had an area of about 70 hectares. An agricultural watershed (W3A) with an area of about 55 hectares was calibrated with the W3B watershed prior to any treatment. Earthen bunds -- or levees -- were constructed along the field boundaries of watershed W3A.

The third watershed (W2C, 4.4 ha) was completely covered with *Shorea robusta* (the main crop) but was used for grazing by cattle. Brushwood check dams were constructed in this watershed to study their influence on soil and water losses.

Arundo domax grass was planted in the gully beds to strengthen the check dams and to rehabilitate the beds with vegetation. Rhizomes also were planted in shallow trenches on the upstream side of the check dams. Cuttings were planted in a total area of about 1,500 square meters.

Data on rainfall, runoff and soil loss from the watersheds were collected from the pretreatment period (1960-69) and the post-treatment period (1972-80). Rainfall was measured by an automatic rain gauge installed in the meteorological observatory in the vicinity of these watersheds. Runoff data were measured using a pre-calibrated triangular weir and an automatic F-type stage level recorder. Runoff samples were analyzed for suspended sediment concentration. At the same time, the bed load was measured in a silt collection tank (or bed load trap) located on the upstream side of the structure.

Regression analyses of peak rates and volumes of runoff before and after treatment produced R^2 values which were relatively low (ranging from 0.4 to 0.85), but the tests confirmed that clearing of natural forests and cultivating them without any soil conservation measures increases the peak rates and overall volume of runoff from such watersheds. *Peak runoff rates* from the agricultural watershed were 72% higher than those from the forest watershed subject to grazing. However, when such an agricultural watershed is treated with bunding, the peak rates can be as little as 14% higher. Similarly, the volume of runoff produced by the agricultural watershed was initially 15% higher than the runoff from the forest watershed. After bunding however, the runoff volume was reduced to 28% of that from the forested watershed.

Runoff recession times of the watersheds, before and after treatment, were observed to be linearly related. The observed relationships were:

$$Yr = 213 + 1.047\,Xr \quad (R^2 = 0.83)\ \text{(before treatment)} \qquad (1)$$
$$Yr = -131 + 0.97\,Xr \quad (R^2 = 0.80)\ \text{(after treatment)} \qquad (2)$$

Where, Yr = recession time (min) for the agricultural watershed (W3A), and Xr = recession time (min) for the control forest watershed (W3B).

Treatment resulted in an observed 7% increase in recession time. Thus, the opportunity for infiltration/ground water discharge increased. The field bunding also had an impact on cropping patterns in the watershed. Following treatment, more than 50% of the watershed area was used to grow transplanted paddy rice, sugarcane and the like, because of the improved moisture regime. Before treatment, the watershed's main crop was corn.

In the forested watershed (W2C), the brushwood check dams were shown to be effective barriers, reducing runoff volumes. No appreciable effects were observed on the peak rates of runoff, however. The relationships for runoff were:

$$Yr = .076 + 1.96.Xro \quad (r^2 = 0.71) \text{ (before treatment)} \qquad (3)$$
$$Y = 3.11 + 1.18.Xro \quad (r^2 = 0.72) \text{ (after treatment)} \qquad (4)$$

Where, y = volume of runoff (mm) from the forest watershed (W2C), and xro = volume of runoff (mm) from the control forest watershed (W3B).

The collected soil loss data showed that the forest watershed (W2C) had relatively high sediment production rates. Construction of check dams in this watershed reduced average annual soil loss from 4.7 tonnes per hectare to 2.8 tonns per hectare, a 54% reduction. Field bunding in the agricultural watershed reduced soil loss from a level of 2.4 tonnes per hectare to 0.2 tonnes per hectare, a 94% reduction.

The findings reported here suggest the following observations. First, the field bunding method was especially effective in reducing runoff. Bunding compared favourably with forest watersheds in this regard. Preservation of soil stock and control of downstream flooding can be markedly improved, then, by bunding. Furthermore, the concomitant improvement in the moisture regime of bunded fields can provide greater flexibility in cropping practices and an opportunity to produce higher-value crops. Second, brushwood catchment dams also offer benefits for both downstream and upstream areas by controlling runoff volume and sediment levels.

Downstream Effects

Illustration 13. Economic Aspects of Sedimentation

The Phewa Tal Catchment in Nepal.[14] In this case, sedimentation was identified as an important downstream effect of erosion. The interest in analyzing the economic impacts of sedimentation was motivated by a decision to construct a dam downstream.

The approach taken by the author represents a good example of how to integrate technical frameworks with economic methods and of how to value the benefits of an erosion control program. The data on erosion rates were taken from secondary sources, while estimates of sedimentation rates were calculated through sampling methods. By sampling existing sediments and analyzing their composition, it was found that (1) the lake bottom, where these sediments were deposited, had at least two meters in depth of clay deposits and (2) the main source of sedimentation was sheet erosion from upstream (that is, 129,400 t/year).

To estimate the amount of sediments that would actually be deposited in the reservoir area, a sediment delivery ratio of .66 was used. Since the size of the catchment area was estimated at 113 km^2, the total amount of sediments expected to be deposited was 86,640 t/year. To analyze the sediment deposition process, three variables were taken into account, sediment delivery ratio (D), trap efficiency of the reservoir (T) and erosion from the catchment (E). The mathematical expression used was,

$$S = E \times D \times T$$

The sediment compaction process was approximated using the following formula,

$$D = D_1 + K \log T \ (0.01602)$$

where D is the density of sediments, D_1 is the density after one year, K is a constant (10.7 for clay) and T is the total number of years. The value of D was one tonne per cubic meter. Thus, given the estimated rate of sedimentation and the volume of the reservoir area (that is, 30.4 x 106) it was calculated that the lake would fill up with sediments at a rate of 0.285% per year, or in 350 years.

The author estimated that by improving the management of upstream land, the economic life of the reservoir could be lengthened from 350 to 400 years. A discount rate of 10% was used. Given these estimates, the major implication is that the value of the discount factors in year 350 to year 400 is so large that benefits during that period are worth practically zero today.[15] Another way to put it is by saying that whatever happens during that period does not influence decision makers today.

Up to this point we have focused on one characteristic of sedimentation (that is, volume) and have related it to the potential value of the asset under consideration (that is, a dam). As shown below, sedimentation has other characteristics (for example, nutrient content) that may be used as a proxy for benefits foregone.

Flemming concluded that approximately 13,800 kg of phosphorus were washed away into the reservoir every year. A detailed analysis of the northern part of the reservoir showed the following losses in nutrients, $P_2 0_5$ = 837 kg/ha (average of 35 sites), N = 6,277 kg/ha (average of 21 sites), K_2 0 = 366 hg/ha (average of 36 sites).

These estimates are very useful to the economist who would like to use the "replacement value approach." Losses in NPK are reflected in lower productivity upstream or in substantially higher expenditures for chemical fertilizers. Because chemical fertilizer is *traded in the market*, estimates of project benefits can be computed via the willingness to pay for those inputs. It would have been interesting to know how much money each farmer spent fertilizers and to compare that with the value of nutrients lost.

Lukkos Watershed in Morocco.[16] The main objective here is to show how economic analysis can be integrated with technical frameworks and to illustrate how the authors used different methods of valuation. Given existing rates of sedimentation, two important effects were identified, (1) water shortages that would affect downstream irrigation and (2) losses in productivity up-stream. The rationale behind this approach was, first, that the volume of sediments (a characteristic) would compete for space with the water stored downstream. Since policy makers were not prepared to sacrifice every other purpose but irrigation, estimation of agricultural benefits foregone (on approximately 25,000 hectares) was an important issue here. And second, that sediments deposited in the reservoir area have a very high nutrient content (another characteristic. Thus, their market value was also used to approximate the monetary value of the project's benefits.

The authors used the Universal Soil Loss Equation (USLE) to estimate the erosion rate in the catchment. The variables used in the equation were the rainfall erosivity factor (400), the soil erodibility factor (0.15), the topographic factor (10) and the vegetation management factor (0.15, with a ground cover of 30%). The latter variable was used as a control factor "with" and "without" the recommended practices. The authors estimated an erosion rate of 90 tonnes/ha.

In addition, it was estimated that the density of sediments was 1.6 tonnes/m^3, or an equivalent of 56.3 m^3/ha/year and a sediment delivery ratio of 0.39. The authors assumed that no major change in the level of sedimentation would take place in the first five years with the project. Therefore, the supply constraint to irrigation downstream would be effect after five years, when they estimated that the dead storage area of the reservoir would be filled up. As stated elsewhere in this book, sediments do not necessarily move into the dead storage area of the reservoir. Thus, economic costs due to sedimentation may appear before the fifth year.

Based on *volume* of sedimentation, the authors estimated that the value of benefits would be equivalent to the value of agricultural output foregone because of shortages of irrigation water. The authors also estimated the *replacement value of nutrients* and related this value to crop yields and productivity. Sampling techniques were used to determine and quantify the chemical composition of sediments. Using economic prices for chemical fertilizers, they estimated a replacement value of US$37 per hectare. Since farmers use different amounts of fertilizer, the estimated replacement value was found to be between 88% and 177% of the actual expenditures for chemical fertilizers.

Potential hydroelectric power, drinking water supply and flood-control related-benefits were not taken into account because of lack of data. The rate of return of the soil conservation project was estimated at 12.1%

Poza Honda Watershed in Ecuador.[17] The Poza Honda is located about 200 km southwest of Quito and covers an area of approximately 175 km^2. A water reservoir of 100 million cubic meters is located in the basin. This reservoir supplies drinking water, as well as irrigation water for 9,500 hectares in the Portoviejo Valley.

Nearly 60% of the watershed has slopes greater than 25%. Aerial photographs taken in 1975 indicated that 55% of the area was natural forests, 6% was in coffee and cocoa cultivation, 22% was grazing land and 14% was farming and grazing land. It was estimated that 75% of the grazing and farming/grazing area was affected by sheet erosion.

The reservoir, which occupied the remaining 3% of the area, had very severe sedimentation problems. The actual rate of sedimentation was approximately 10 times higher than the assumed rate at appraisal. There was a deposition rate of 4% per year, giving the reservoir an economic life of only 25 years rather than the 50 years assumed during construction.

The erosion rate in areas not under forest was estimated at a very high 458 m^3/ha/year. This was equivalent to a soil depletion rate of 4.6 cm/year. The erosion rate estimate was based on sedimentation studies.

A development program including both mechanical and biological soil conservation methods was drawn up for the watershed. Since this case study focuses only on sedimentation, no computational details are given here.

The value of costs of the conservation program was estimated at 45.6 million sucres. This covered the cost considered necessary to extend the life of the reservoir from 25 to 50 years. A 50 year project with a conservation program was calculated to have net benefits of 456 million sucres, while a 25 year project without the conservation program had negative net benefits of 312 million sucres. The difference between the net benefits of the two projects was 768 million sucres, equivalent to US$ 31 million. The conservation program under consideration, costing $1.8 million, was expected to result in project benefits in excess of $30 million.

Illustration 14. Reservoir Sedimentation[18]

A review was made of the probable sediment inflow and deposition in a proposed reservoir above the Yonki Dam site on the Ramu River (Papua, New Guinea). The steep slopes of the upland area in question was subject to erosion by landslides and intensive runoff during heavy rainfall. Mining operations were also an important agent of erosion.

The following factors were assumed to affect sediment yields, (1) rainfall amount and intensity, (2) soil type and geologic formation, (3) ground cover, (4) land use, (5) topography, (6) upland erosion, (7) water runoff, (8) sediment characteristics (for example, grain size, mineral composition) and (9) channel hydraulic characteristics.

The long-term average annual sediment yield for the Ramu River was estimated from the following equation:

$$Qs = 3901A^{-0.24}$$

Where, Qs = sediment yield ($m^3/km^2/yr$), A = drainage area (km^2).

The equation gave a yield estimate of 780 $m^3/km^2/yr$. Based on these consumptions, the average annual sediment load was estimated at 585,200 tonnes per year.

The next step in estimating the total sediment load required the computation of the bed load, including unmeasured load at the sampling site. It was concluded that the bed load as a percentage of the suspended load would be 10%. This resulted in an average annual bed load of 58,520 tonnes per year, or a total sediment load of 643,720 tonnes per year.

Given the storage capacity-inflow relationship for the Yonki Reservoir of -- capacity = 335 Mm^3, Inflow = 951 Mm^3 -- and using a trap efficiency curve developed by Brunne, a trap efficiency of 97% was estimated. This implied that 624,000 tonnes per year of sediments would be deposited in the reservoir.

It was assumed that certain factors would affect the density of the sediments deposited in the reservoir, (1) the manner in which the reservoir was operated, (2) the texture and size of deposited sediment particles, and (3) the compaction or consolidation rate of the deposited sediments. The following estimates were made for a poorly operated reservoir, For 48 years, 1198 kg/m_3, for 100 years, 1231 kg/m^3.

In many of the computations considered earlier, it was assumed that sediments will be deposited in the dead storage area of a reservoir. This is not always the case. Detailed studies like this one show that the dynamics of the sediment deposition process are more complex. The method used to assess the distribution of sediment in the reservoir consisted of adjusting the original area and capacity curves throughout the reservoir depth. This distribution required data on contour areas planimetered from a topographic map to compute storage capacity.

The next step consisted of computing the elevation of sediment deposited at the dam. In 48 years, the elevation will be 1220 meters, or 5 meters higher than the reservoir bottom. By the same computation method, the 100-year estimate for sediment elevation is 1223 meters.

Finally, a delta profile was constructed to assess the impact of deposits of sand-size or coarser material in the upper reaches of the reservoir. Such deposits may cause a rise in elevations of backwater, possibly inundating farmlands adjacent to those sections of the reservoir. The delta profile was determined by defining the location of the delta's upper limit, the relevant slopes and the 48-year sand volume. In this case, the steep slope above the reservoir would limit the deposits to an elevation below 1258 meters.

Illustration 15. Benefit Cost Analysis of Hydrometeorlogical Data[19]

The main objectives here are to discuss the need and relative merits of data collection and processing activities and to illustrate how BCA has been applied to determine "how much data is enough" for different analytical purposes.

Development projects activities require investments in infrastructure, manpower, hardware and software. In a sense, these investments are similar to those in other sectors of the economy. The size of these investments and their benefits and costs, depend on their end uses and on the quantity of data required to justify them.

The UN (1977) estimated that the total cost of obtaining the data needed in hydrometeorology (the network of measuring stations and related services) was US$1.3 billion (in 1976 prices). This worldwide estimate did not include the Peoples Republic of China, the Soviet Union, or Canada.[20]

WMO (1982) estimated that the cost of bringing the networks of hydrological stations in Africa to a minimum standard was close to US$36 million (also in 1976 prices). This estimate included discharge, non-recording rain gauges, recording rain gauges, evaporation stations and data processing, analysis and interpretation.

Alternative Approaches. Several things are worth mentioning before the alternatives are presented. First, data systems in this field can be classified into four groups, (1) inventories of water resources -- this data is shared by a large number of users and is intended to reduce errors in existing information, (2) data on planning and design -- this information is mostly used to determine the proper size and characteristics of proposed reservoirs or other investments in flood control, (3) data on operations, and (4) forecasting data -- these data are needed to determine future sedimentation rates, flood frequency, design floods and rainfall patterns.

Second, the extent to which policy makers will accept the results of a benefit/cost analysis will depend on the creditability of benefit estimates. These estimates will determine how much data must be collected. These data will vary, depending upon existing regulations and standards (for example, water quality standards).

Although the alternatives demonstrate the use of the economic criterion, it is important to say a few words about the statistical criterion.

Any model designed to forecast a hydrological phenomenon using econometric methods will need a minimum set of data to achieve a "good" predicting value. It is often said that a forecasting model needs at least 25 observations of a variable that is distributed in a "bell-shaped," or "normal" curve. In the case of time-series analysis, the model will need at least 12 observations to achieve an acceptable level or predicability.

To avoid major mistakes, the appraisal team will need to know a great deal about the natural phenomenon under analysis. This will be particularly important for phenomena that behave in cyclical fashion, since the observations at hand may reflect only a part of the cycle.

The main point to keep in mind at this stage is that the question "how much data is enough" can only be answered by using a statistical criterion. As stated earlier, this criterion will often be modified by existing regulations and standards as well as by economic considerations.

The Use of Economics in Hydrometric Data. In assessing the economic merits of systems for the collection of data about hydrological phenomena, three approaches have been used. The first is *apportionment of benefits* from water resource projects. This approach considers the costs associated with data collection, processing and use as part of the total cost of the project. The benefits of data collection are computed by assigning a portion of total project benefits to the data component. This method is simple and straightforward. However, it has several disadvantages, the most important being, a) lack of a scientific method of attributing benefits to this component and b) inability to give any indication of incremental benefits that would result from improving the data system.

The second approach is *error reduction in the estimates*. This approach assumes that there is a strong correlation between the benefits of an increase in the quantity or quality of hydrological information and the standard error of the estimate, which in turn affects the parametric value under question. The "expansion" in the data base could take several forms, such as an increase in the frequency of measurement, in the number of hydrometeorlogical stations, in the number of years of observation, or improvement of interpolation techniques. All of these alternatives will have a specific impact on costs.

A third approach is *error surrogate techniques*. This approach consists of maximizing the decrease in the standard error, given budget constraints. This approach is particularly useful when the relationship between the benefits of increased hydrological information and the percent of the standard error is difficult to define. Deterministic as well as probabilistic methods are used to carry out individual evaluations.

H.J. Day's Case.[21] The essence of the procedure presented here can be summarized by the following statements, (1) evaluation without any forecasting service, (2) evaluation with a forecasting service and (3) assessment based on (1) and (2) taking into account a reliability factor of hydrological analysis in forecasting.

One application of this method is *flood forecasting*. As is already known, the objective of a flood control project is to reduce the probability of potential damage. Three key elements will be of importance to decision-makers, length of warning time, magnitude of reducible damage and loss of life and efficiency of response to warning. The magnitude of property damage and loss of life is particularly sensitive to length of warning time (the period from release of a flood forecast to its actual occurrence). The term "reducible damage" is important because there will always be damages that cannot be avoided even if the length of the warning time is increased (for example, damage to farm fences). Efficiency in response to warning depends upon the time of the day the warning is received, the time elapsed since the last flood and the accuracy of past forecasts.

The model used here specifies a flood stage-damage curve for different types of structures (for example, housing).[22] The efficiency of flood warning can be estimated by relating flood depth (or crest elevation) to the water level in flooded structures and using the *stage-damage* curve for each structure. The following summarizes the problem.

$$E\ (D) = \Sigma^{n}{}_{i=1}\ P_i\ .\ D_i$$

Where E(D) is the expected average annual loss, n is the number of years, P^i is the probability of a flood within the flood plain contour interval i-1 to i and Di is the community damage associated with flood reaching to top of step i and a particular warning time.

Two values for E(D) can be calculated, one where there is "no warning," the other where there is "warning." The difference between these values will be the expected annual benefit.

Basso's Case.[23] A benefit-cost analysis of a flood-forecasting system in the Lower Lampa River in El Salvador was carried out. The system consists of a 10cm radar, 6 pluviometric stations, 2 water-level recorders and transmitting stations to the forecasting center. The capital cost was estimated at C$750,000 and the annual cost at C$160,000. Flood damages were assessed over a ten-year period during which seven major floods occurred, causing the deaths of 276 people and injuring another 425. The following items were included:

1. Loss in human life or injury (the "revenue foregone" method).
2. Loss of cattle.
3. Loss or damage to houses or other properties.
4. Damage to roads.
5. Agriculture-related losses.

The benefit of the forecasting system was equated to avoidable loss, which amounted to C$8.8 million. The benefit-cost ratio was 4.5:1. The benefits were probably overestimated, since, as noted in Day's case, certain damages are unavoidable despite the existence of a forecast system.

M.A. Omar's Case.[24] Here the author estimated the value of meteorological information in relation to agricultural sector management. Weather forecasts help farmers decide when to plant, what seeds to use, when to harvest a particular crop, when to invest in infrastructure and the like.

Omar gives several examples, most of which were related to the selection of suitable times for agricultural operations. One study in the U.K. showed that the proper use of weather forecasts to choose when to cut and dry hay could increase yield and quality by 10%, thereby increasing winter milk production by 2%. It was estimated that the annual value of the weather forecasting service was at least Two million pounds. In a pilot project in the United States, it was found that an adequate weather forecasting system would enable farmers to avoid premature planting, carry out efficient pest control and enhance other farm operations. The cost of the pilot project was US $5,000 and the benefit/cost ratio was about 40:1.

In general, studies in both developed and developing countries have found extremely high benefit-cost ratios in this setting.

Illustration 16. The Effect of Land Use on Flood Flow Peaks[25]

In an effort to improve flood estimation methods in East Africa, the United Kingdom's Transport and Road Research Laboratory assisted the governments of Kenya and Uganda in a data collection and analysis program which concentrated on factors connected with the flooding of small river basins. Data were collected from 1966 to 1973 from 14 rural basins and 5 urban basis. The rural basins were chosen to cover the factors that determine flood runoff, rainfall, land use and topography, the last being linked to soil type. Most of the basins were small (up to 14 km^2), in order to facilitate measurement of rainfall volumes and flood flow. The urban basins were considerably smaller than the rural ones and were selected to provide a range of soil types, slopes and types of development.

Due to the unreliability of hydrographic data on these basins and the relatively small number of high-intensity storms that occur in them, the development of a general flood production model from rainfall-runoff correlations was deemed inadvisable. Instead, with the four years of data available from each basin, runoff was simulated using a linear reservoir analogue and the concept of contributing area.

In the *rural* flood model, each basin was divided into a number of sub-basins. The sub-basin model may be summarized as follows:

a) Early rain fills the initial retention area (Y). Runoff at this stage is zero.
b) Subsequent rain falling on the parts of the basin from which runoff occurs (contributing area, or C) enters reservoir storage (S).
c) Runoff is then given by $q = S/K$, where K is reservoir lag time.

In order to translate runoff down and stream system to the outfall, a finite difference technique was applied for each large storm in a basin and for a variety of values for the parameters C, Y and K. Recorded and predicted hydrographs were then compared and the optimum values of parameters arrived at by goodness-of-fit test. To develop a general flood model, the variations in these optimum values among the basins were examined.

The four characteristics of the basins considered in the analysis were: soil type, slope, type of vegetation or land use (particularly in valley bottoms) and basins wetness. Standard coefficients were derived from earlier studies of soil type/slope zones in each basin. Similarly, wetness and land use factors were derived by classifying sub-basins into antecedent wetness zones and by comparing recorded volumes of runoff in various land-use areas with those that would have occurred in a standard grassed basin. The design value of the contributing area coefficient is therefore given by:

$$C = C_s C_w C_l$$

Where, C_s is the standard value of the contributing area coefficient for a grassed basin at field capacity, C_w is the wetness factor and C_l is the land use factor. The total volume of runoff is given by:

$$RO = (P\text{-}Y)\ C_A\ A103\ m_3$$

Where, P is the storm rainfall in mm during the time period equal to the base time, Y is the initial retention in mm, C is the contributing area coefficient and A is the basin area in km^2.

If the hydrograph base time is measured to a point on the recession curve at which the flow is one-tenth of the peak flow, then the volume under the hydrograph is approximately 7% less than the total runoff given by equation (2). The average flow Q^* is therefore given by:

$$Q^* = \frac{0.92\ RO}{3600\ T}$$

Where, T is the hydrograph base time in hours. The base time is estimated from the simulation exercise by using: (a) the rainfall time, (b) the recession time for surface flow and (c) the attenuation of the flood wave in the stream system.

Earlier studies have shown a quantitative relationship (peak flow factor F) between average flow and peak flow. For very short reservoir lag times (K= 0.2 hours), F was 2.8 ∓ 10%. For lag times longer than one hour, F was 2.3 ∓ 10%. By relating the peak flow factor to average flow, as in the following equation, the value for peak flow Q is estimated,

$$Q = Q^*\ (F)$$

The approach adopted for the *urban* sub-basins was identical to that used for the sub-catchment modelling in the rural flood model: The flow from a contributing area C was routed through a linear reservoir with lag time K. The urban values for C were increased by approximately 50% over the rural values to allow for the much better drainage systems found in urban areas.

For rural basins in this study, the volume of runoff was directly proportional to the land use factor. Peak flow depended on volume of runoff as well as lag time K, which in turn was altered when land use changed. For example, the estimate of K for arid basins was 0.1 hour, while that for papyrus swamp valley bottoms was 20.0 hours.

The change in peak flow was more difficult to predict for urbanised basins. Virtually all of the rain falling on paved, sewer-connected areas will quickly enter sewers. The drainage density of the sewer network is likely to be much greater than was the case before urbanization. Runoff from unpaved areas therefore can also be expected to be greater and to occur more quickly. Due to the reduced hydraulic roughness of sewer systems as compared to natural streams, flow velocities are likely to be greater in urban systems. Since both volume and speed of runoff increase in urban areas, peak flows tend to be greater.

To explore the effect of land use changes on peak flows, the two models were applied in a simulation of runoff in four larger urban basins in two-year recurrence-interval design storms. Three land uses were compared, forest, grassland and urban. Especially large increases in peak flow (approximately ten times) resulted from a change in use from forest to grassland. Further increases in peak flow resulting from urbanization were much more variable, fluctuating from two to five times more in the basins studied. Several things may be noted with regard to the economic implications of this example:

a) Land use change may be a central theme of project activity, as in a rural development scheme which emphasizes cropping innovations. Alternatively, land use change may be a secondary effect of project work (for example, an agricultural credit project which induces pastoralists to enter into crop production). But whether land use is central or peripheral to project objectives, the consequences of land use change on flood vulnerability -- locally and downstream of a project area -- are often substantial.
b) The types of economic activity in a vulnerable region will suggest critical variables worthy of careful technical-economic coordination, for example, the *velocity* of a flood is especially relevant in assessing the risk of damage to infrastructure, such as roads and buildings. While velocity also needs to be considered in relation to rural areas, velocity is especially important for risk analysis for urban areas. The time period and area of inundation, on the other hand, may be variables of greater importance in agricultural areas.

c) Income foregone measures are likely to rely on different types of data, depending on the predominant types of economic activity in the area analyzed. In agricultural settings, the measurement commonly is agricultural yield. The value of residence, work sites and infrastructure are reasonable bases for assessing damage risk in urban areas.

d) In economic valuation of flood damage, it may be appropriate to construct individual damage functions for each affected sub-basin. In the case of a single affected basin, multiple damage functions may still be required in order to account for variation in flood types across seasons.

e) Alternative flood prevention and control measures need to be considered in light of the flood variety discussed above. Special attention should be paid to adaptation of general flood control plans to the requirements of local sub-basins.

Illustration 17. Estimation of Flood Control Benefits

Several studies on estimating flood control benefits have been carried out. Some of the examples used here focus on residential rather than agriculture-related damages. This should not pose a problem, since the main objective here is to illustrate methodology and the actual analytical framework applied.

Using Implicit Price Equations.[26] The following case shows an application of the *land value approach*. As stated earlier, many environmental factors are not traded in the market (pollution, floods) but changes can alter the characteristics of assets that are traded. A proxy value for flood damages is obtained by assessing corresponding changes in land values. This approach explicitly identifies the characteristics of land as an asset. These characteristics are assumed to be reflected in land prices.

The land value approach assumes that consumers will buy and sell land in the market at a price which reflects the levels of these characteristics. In other words, the final price of any given parcel of land will be an aggregation of implicit prices for those characteristics.

This method of determining the final price is also common to agricultural lands, although research that would correlate these characteristics with land values is not often carried out in developing countries.

Written in a functional form, one can say that land values in this case of residential land (R) are determined by a bundle of characteristics (Ci):

$$V(R) = V(C_1, C_2 \ldots\ldots, C_n)$$

Econometric methods are used to determine the value of V(R). The application presented here focused on the city of Sutherlin, Oregon, with a population of 4,560 in 1980. Flooding was a common natural phenomenon in Sutherlin where, on average, downtown as well as residential areas were inundated once every year.

Four clusters of characteristics were defined, physical characteristics of the land (for example, lot size), accessibility (for example, location of centers of employment and other centers), environmental quality (for example, air quality, flood hazards) and availability of public sector services. As stated earlier, flood damages can be mitigated, depending on the type of "design flood" for which investments are made. The net reduction in flood damages is estimated by subtracting remaining damages "with" and "without" the project. These were calculated by estimating an implicit *price equation* for each case. A dummy variable was used to reflect flood damage with and without the project.[27]

The net reduction is flood damages was estimated at $735 per lot (that is, $1,600 without and $865 with the project. The equation for the "without" and "with" project situation (for a sample of 235 lots) was:

$$V(R) = -28.65 + 0.05L - 0.14D_1 - 0.03D_2 + 3.64D_3 + 0.02B^* - 200F^*$$

$$V(R) = 3886 + 0.44L^* - 3.9D_1 + 0.001D_1^2 + 0.36D^2 + 0.00003D_2^2 - 0.05B + 194NR - 865F^*$$

Where * denotes a significance level of 90% or greater. And where, L = lot size, D_1 & D_2 = distance variable, B = value of improvements, NR = nonresidential land use (dummy), F = Floods (dummy)

Again, the basic assumption of the study was that changes in the characteristics of the environment (in this case, the degree of susceptibility to floods) are capitalized in the value of land.

Income Foregone from Crop Production. Most of the literature on estimating flood damage to agriculture applies the "income foregone approach." As stated earlier, the object of this approach is to estimate the value of agricultural production lost due to floods. This approach has been shown to be useful within a developing country context, where land market transactions place a higher priority on other characteristics than the land's environmental attributes. In such a case, the value of income foregone is a good proxy for benefits.

The method consists of dividing a flood plain into relatively homogeneous segments (that is, reaches) and then estimating benefits from reduction in flood damages. The computation of these damages is related to crop losses, which in turn depend upon the frequency, the depth and the total time of a given flood. In this case, the hydrologist of the appraisal team prepares a table showing the number of acres flooded to different depths by floods of various depths or stages. Based on this data, a stage-damage table is put together showing crop damage for different types of floods for each month in any given year.

Two short-cut methods may be applied. *Method I* approaches the problem from an income standpoint, where aggregate estimates of average yields "with" and "without" the project are compared. Therefore, the analyst will have to estimate what net income would be if no floods occur in the area and compare this estimate to an estimate of net income if floods do occur.

Method II starts by cataloguing the number of floods -- over, say, a 20 year period -- for each quarter of the year. Next the hydrologist will furnish data on the acres inundated at different stages for varying degrees of flooding. One is then ready to relate these data to different crops on a quarterly and depth of inundation basis. Using yields and prices, a total value of damages is estimated.

Integrating Technical and Economic Frameworks.[28] This case shows how a relatively simple technical framework can easily be integrated with traditional benefit-cost analysis techniques.

The project studied was located in the Lower Cimanuk Basin along the north coast of West Java, Indonesia. Monsoon rains occur between the months of November and May and floods may occur within six hours of heavy downpour. Floods generally reach the vulnerable stretches of the Lower Cimanuk, in the early hours of the morning, when communications and emergency repairs to levees are harder to make.

The Cimanuk River and six smaller streams end in the lower part of the basin. Channel capacity diminishes toward the outlet and thus these rivers frequently overflow their banks, causing extensive damage to property, infrastructure and crops in this fertile and densely populated basin (that is, 1.7 million people).

Flood flows of the Cimanuk River at Rentag Weir were derived from a frequency analysis of recorded floods (17-30 years) at two stations. The results of these analyses were checked by synthetic methods, using storm patterns derived from an analysis of extreme precipitation as well as exceptional storms recorded in the area over a 41 year period. It was found that the height of the 40 year flood is little different from that of the 20 year flood. Project works were therefore sized to accommodate the 25 year flood.

The total cost of the project was estimated at US$77.0 million, with a base cost (for civil works, equipment and services) of US$52.4 million.

Flooded areas can be distinguished from rainfall-inundated areas and floods can be traced back to one or several breaches in the existing levees. Studies were conducted to observe the frequency distribution of floods over the period 1970-77. Given the very high correlation between the theoretical probability distribution used and the results given by the model, it was concluded that the 1970-77 data were presentative of that frequency distribution. On the basis of yearly data, the correspondence between maximum floods and areas inundated could be studied. Based on the data available, a very poor correlation was found between the flooded areas and flood discharge.

This lack of correlation was related to the fact that flooding was not due to overtopping of levees in slowly rising floods but to breaches in the levees. The mission team concluded that floods in the Lower Cimanuk basin are largely random. Consequently, the randomness associated with the extent of flooding led to randomness in the value of damages. This randomness was compounded by variations in damage per unit area.

Apart from the breaching of levees, floods commonly damaged crops, houses, household effects and animals. Only on low ground areas did the water tend to remain for more than three days. This implied that damages were largely due to flowing water instead of water deposition and were likely to be greater if the flood occurred at night, when farmers found it harder to take preventive measures.

Crop damages took two forms, depending on the month in which the floods occurred. If the flood took place late in the wet season, advanced paddy crops inundated by flood waters suffered yield reductions which varied from negligible to total, depending on the depth and duration of the inundation, the velocity and turbidity of the water and the age of the crop. The average estimated yield loss was 25%. If flooding occurred earlier in the wet season, the immature rice crops were completely destroyed. In that case however, there was still time for farmers to replant their crops, but paddy yields were reduced by about 5%. These losses, plus the sunk costs of inputs and labor of the destroyed crop, were estimated at Rp 125,400 per hectare.

Damages to houses and household effects were estimated from actual counts of inundated houses and appraisals of the value of household effects over the period 1970-77. The average damage was estimated at Rp

180,000 per household (50% of that was the value of household effects lost).

Floods also disrupted transportation. Nineteen km of kabupaten roads and thirteen km of provincial roads were affected by floods. Roads, canals and bridges occasionally were washed away, but the most conspicuous damage was to road surfaces. The benefits of the project were quantified by estimating a value for *mitigation costs*. The average road repair cost was about Rp. 16 million per km, while repairs to other structures were about Rp. 105,000 per inundated hectare. Although it was also recognized that there were other forms of flood damage (for example, health hazards, loss of human lives), the data was insufficient to quantify these costs.

By aggregating benefits (the costs of income otherwise foregone), the appraisal teams estimated a project rate of return of 16%. The economic value of expected flood damages was estimated at Rp. 5.3 billion.

Illustration 18. Changes in Water Quality

Many factors affect water quality, but in the last analysis water quality depends on the water's end use. Water may be of "good" quality for use in irrigation, but of "bad" quality for fisheries development.

There are several *direct methods* of quantifying water quality, most of which are based on predefined acceptable levels of chemical (or other) substances in the water. If these standards are accepted, economic analysis simply involves a determination of the minimum cost of achieving those levels. If the goal however, is to raise water quality above merely acceptable levels, economic analysis then involves addressing both benefits and costs.

When water is used for such purposes as recreation, there are *indirect methods* of assessing water quality -- for example, by assessing public willingness to pay for recreation. The assumption is that people will be wiling to pay more for cleaner water than for water of lesser quality (that is, turbid water).[29] Two indirect methods are widely used, interviews with users of water facilities (for example, lakes, rivers, dams) and assessments of the cost of travelling to a facility.

Comparison of Objective and Subjective Ratings.[30] This method involves comparisons of user's subjective perceptions of water quality with some sort of objective rating. In the case illustrated here, a Lake Condition Index (LCI) was used to obtain an objective rating. This value of benefits was then compared with the cost of a proposed storm sewer diversion project, which was found to be a profitable investment.

A theoretical model was developed assuming that individual satisfaction or utility (U) can be represented as a function of consumption (C) and environmental services (Y),

$$U = U(C, Y)$$

The next stage of the analysis focuses on the degree of complementarity between consumption and services. Most often, analysts assume "weak" complementarity. This means that consumption of a private good is a necessary condition for enjoying a given level of environmental quality. Under such conditions, it is possible to derive the benefits or costs of a quality change in the public good (for example, water quality) from information on the demand for the private good (C).

The next stage of the analysis consists of determining the demand-price relationship for environmental changes by using a proxy variable. In cases where water quality affects the value of recreation, most analysts have used as a proxy variable "recreation visits (RV)." The number of recreation visits will change with changes in the level of water quality with the key question being how much the consumer will be willing to pay for the change in water quality.

The assessment can proceed by following three steps, (1) assessing a change in price from Po to P1. Given demand curve Do, the consumer must be compensated by consumer surplus so that they are not made worse off by this change in price, (2) assessing a change in water quality (WQ). Given the assumption of weak complementarity, the implicit hypothesis is that the consumer's utility remains practically unaffected and no compensation is needed, and (3) assessing a change in price to Po. If consumers are willing to pay the equivalent of the new consumer surplus, the new result is the difference between the consumer surplus before and after the water quality change.

Consumer willingness to pay can be estimated through direct interviews. Thus, a commonly used method is to estimate the cost of travelling to the facility, *the travel-cost approach*. For the interview procedure to be useful however, consumers must have some perception of the change in water quality.

Scientists can obtain a more precise notion of changes in water quality by making precise measurements of changes in such things as turbidity, dissolved oxygen and BOD. As stated earlier, the Lake Condition Index (LCI) was used in this case to estimate the relationship between "objective" and "subjective" measurements of water quality.

The LCI was developed to classify all lakes in the state of Wisconsin larger than 100 acres. A point system for each component of the index was used. The most important components of the index are dissolved oxygen, water transparency, fish winterkill and the extent of algae or macrophyte growth. The level of water quality was measured according to a ranking system, 0 being the highest level of quality and 23 the lowest.

To test the relationship between the objective (LCI) and the subjective water quality ratings, several interviews were conducted at each lake. The respondents were asked to rank the lakes water quality from 0 to 23. Then a test was run to measure the effectiveness of the LCI in predicting respondents' perception of water quality. This was done by regressing the average rating (R) of all respondents at each lake on the corresponding LCI for that lake. The results were as follows,

$$L_N R = 1.948 + \underset{(3.37)}{0.364\,LCI} \qquad R^2 = 0.694 \quad n = 7.$$

Using the above mentioned equation, a model to estimate the benefits of water quality was developed,

$$V_{ij} = a_j + b_{ij}/X_{ijk} + e_{ij}$$

Where, V_{ij} = the number of visits of consumer i to lake j, b_{ij} = parameters to be estimated, a_j = parameter to be estimated, X_{ijk} = is the value of the variable for consumer i to lake j, and e_{ij} = the error term.

The estimated demand curve was,

$$Vo = 43.22 - \underset{(.950)}{0.317C} + \underset{(2.10)}{0.008C^2} - \underset{(2.93)}{5.264LnR} - \underset{(3.63)}{0.162T} + \underset{(2.81)}{0.0003T^2} - \underset{(2.35)}{0.321I} \qquad R2 = .203 \quad N = 195$$

Where, Vo is the number of visits for the year, C is the total variable cost per trip, R is the consumer's rating of water quality, T is the round trip time, I is the consumer's annual income.

The total cost was estimated at \$429,038, with the estimated annual benefit was \$38,964 which, at a 10% discount rate over 20 years, is \$331,740. Since the cost of the water quality project was estimated at \$175,000, it was concluded that it would appear to be a wise decision for water resource managers to recommend the project.

Notes

1. Drawn from Gunter Shramm (1982).

2. Drawn from Johnson and Kolavalli (1984), pp. 185-188.

3. This case is based on Young and Walker (1982).

4. From Rutherberg and Lehmann (1980), pp 300-303.

5. Based on Walker (1982).

6. This case is based on Elwell and Stocking (1984), pp. 148-150.

7. Based on Jose and Markland (1981), pp. 114-116.

8. Based on Kim and Dixon (1984).

9. Based on Elkins (1981) pp. 12-14 and Bauder (1981), pp. 15-17.

10. This case is based on Mitchell *et al.* (1980), pp. 233-236.

11. Based on Gupta *et al.* (1973), pp. 205-211.

12. Based on Mita (no date).

13. Based on Sastry and Narayana (1984), pp. 14-21.

14. Based on Flemming (1983).

15. Flemming's paper includes other benefits and costs which we have not taken into account. "Sedimentation" is used here to illustrate certain points. Thus, the conclusions here should not be attributed to Flemming.

16. The material was drawn from Brooks *et al.* (1982).

17. Based on Flemming (1979).

18. Drawn from Pemberton (1983).

19. The material was drawn from WMO (1982), Day (1973), WMO (not dated), UN/UNESCO (1977), UN/UNESCO (1984), WMO (1977), Dyhr-Nielsen (1982), and Omar (1980).

20. This estimate does not include the cost of data related activities for groundwater assessment.

21. Taken from H. J. Day (1973).

22. This model may also apply to agriculture when crop-related damage curves can be drawn based on how different crops respond to the level of water and the duration of the flood.

23. Taken from H. J. Day (1973).

24. Based on Omar (1980). This article contains valuable information on methods of economic evaluation by other scientists.

25. Based on Fiddles (1980).

26. Based on Thompson and Stoevener (1983), pp. 889-895.

27. This case is based on Freund and Tolley (1979).

28. This case is based on World Bank, Indonesia-Lower Cimanuk Basin Flood Control Project. Staff Appraisal Report No. 23306-IND. April 2, 1979.

29. Here, we assume that soil erosion affects water quality. Water is an input to the production of a tradeable commodity (e.g., recreation). However, the value that an individual gives to recreation depends on several "characteristics," one of which is water quality.

30. This case is based on Bouwes and Schneider (1978).

PART FOUR

12

Project Appraisal Issues

Technical Issues

The technical issues involved in soil conservation projects are diverse. Most of them are related to the multi-dimensional characteristics of soil erosion, the difficulties in establishing a strong relationship between erosion and changes in productivity, the nonuniformity of the soil erosion process, the delivery of sediments to land and water, the episodic nature of the weather, the efficiency of soil conservation practices and the availability of technical data.

Measuring Effects

A major technical issue is that of defining the proper numeraire in which to measure soil erosion as it relates to productivity. Often, the only factor considered to directly affect productivity is the soil erosion rate -- that is, the volume of soil that is detached and transported from the farm. When it comes to quantifying changes in productivity, however the amount of soil that is left on the ground and its characteristics are more important than how much soil is gone. Other factors affecting productivity are the effects of erosion on soil textures, soil nutrient content, the depth of root zone and soil-water retention. Therefore, a simple measure of tonnes per hectare per year or centimetres of topsoil lost for any given length of time are not particularly helpful for economic evaluation.

The same issue is present in the case of projects that have off-farm or downstream effects. Technical measures must include the quantification of parameters that are directly linked to the economic productivity of activities taking place outside the project area. Examples are the chemical content of water in relation to acceptable levels of concentration, a reservoir's usable capacity in relation to its economic life.

Erosion Rates and Crop Yields

Where projects are designed to change the rate at which soils are being eroded over time, analysts face the problem of quantifying an acceptable relationship between agricultural productivity and changes in soil characteristics. This issue is not very well understood, even in highly studied areas of the world and is complicated by the dynamic nature of the soil erosion process, it may take a long time before one perceives (from eyesight) the actual effects of erosion on productivity. In many instances, farmers mask these effects by applying fertilizer.

Moreover, changes in productivity often seem to occur sharply only after soil has eroded to a critical level. This process occurs rather gradually and it is difficult to predict when the critical level will be reached. Even when it is predictable, there are many counteracting factors. Therefore, an understanding of erosion in a static framework is not sufficient.

The quantification of certain downstream effects presents similar problems. Here, one tries to relate such factors as suspended sediments or losses in nutrients (which are transported downstream) to the marginal value of water, which, in turn, represents an input to other economic activities. Thus, a knowledge of the overall effects on productivity is essential in determining which effects are important from an economic standpoint. Unfortunately, current knowledge is inadequate.

Non-Uniformity of Effects

As stated earlier, the rate of soil erosion varies because of changes in such factors as rainfall, soil structure, slope, slope length, vegetation cover, crop management and farming practices. There is great heterogeneity in the way these factors are distributed geographically. The major technical issue is how best to subdivide an area (for example, by types of farms, by watershed or by river basin) in order to determine with some degree of confidence the overall effects of erosion. This is often referred to as the "unit of account" problem.

A high degree of heterogeneity also characterizes the factors that effect downstream effects. In analyzing the major impacts of any given flood, for example, one must define the areas affected by type and nature of flood (for example, frequency, depth, time) and then estimate the size of the areas affected to determine potential benefits and the way in which these benefits can be aggregated. Thus, one needs to subdivide areas in a way that is meaningful for economic analysis.

Sediments Deposition on Land

It is pointed out earlier that the erosion process includes not only the detachment of soil particles but also the transportation of them to another place. It is therefore important to determine how and where the detached soil is likely to be deposited. In many instances, soil moves from one place but remains on-farm or within the watershed. Therefore, one needs to quantify the effects of soil erosion on land losing soil (that is, a negative impact) as well as on land gaining soil (that is, a positive impact). The major technical problems involve establishing how soil is redistributed across affected land areas and how soil deposition affects land productivity.

Erosion Impacts on Water

Soil transported by water changes the quality of that water by adding nutrients and chemicals from the detached soil to the water. The major technical issues here are estimating the relationship between changes in water quality and its value for different kinds of uses and how the quality of water is affected by spatial and intertemporal factors. When water is used for downstream irrigation for instance, a high nutrient concentration may have a positive impact on productivity. However, when water from the same source is used for drinking purposes, public authorities may be required to increase the level of treatment before it is usable at the household level.

Episodic Nature of Erosion

Most data on soil erosion portrays an "average" during a particular period of time. Statistics on such things as "average tonnes of topsoil lost in one year" are plentiful in the literature. Indicators of this type are often misleading however, since the factors affecting the rate of erosion do not behave in a uniform way. Their variability, in turn, causes great variability in the way soils erode. Erosion is basically stimulated by either rain or wind storms. When such storms are intense, they can increase the amount of topsoil losses several fold in a very short time. In some cases, a very large percentage of the total erosion estimated in any particular year occurs during one or a few storms lasting only a few days, or during rains limited to a period of several weeks. An understanding of the episodic nature of the elements that affect erosion is critical to understanding erosion's effects and the ways in which soil erosion programs should be designed.

The major issue for economic evaluation, given the episodic character of soil erosion, is determining the timing of erosion and the corresponding effects of yields and productivity. Flooding provides a good illustration of the problem. Flood damages are more severe if they occur when lands are being cropped. The solution of this technical problem should therefore be framed in a way that includes the "time dimension" of random factors.

Effectiveness of Soil Conservation Practices

Another major technical issue is how to predict the effectiveness of soil conservation practices in bringing the soil to an acceptable level of productivity. Due to the many factors affecting erosion rates, directly and indirectly, there is no simple way to define the effectiveness of soil erosion control practices. It is important to determine how site-specific modifications should be adapted to a particular project and how such modifications are related to productivity, given the current inadequacy of physical measurements of the effects of erosion on and off farm activities.

Factors Which Mask Erosion Effects

Agriculture in developing countries has been stimulated in recent years by the so-called "green revolution." Great increases in yields occurred because of the adoption of high-yielding varieties (HYVs) and more intensive use of water, fertilizers and agrochemicals. Along with the revolution however, has come severe degradation of cultivated land. Because of the growth in yields, soil degradation has not been perceived to be a serious problem. A major issue then, is how to focus attention on the soil degradation process despite great improvements in productivity. These productivity gains mask the effects of erosion up to the point where farmers perceive erosion's negative consequences only after irreversible damages have occurred.

In a recent article, R. Sharma (1983) addresses the issue of losses in micro-nutrients due to the use of HYVs in India. He states that "extensive agriculture based on chemical fertilizers and high yielding varieties has certainly helped to fill many stomachs but agricultural scientists are now discovering to their dismay that it is also depleting the soil, which can lead to a serious fall in crop yields (p. 15)." In the Punjab, India, areas with the highest yields also show the greatest deficiencies in plant micro-nutrients. A typical rice-wheat rotation, yielding nearly 9 tonnes/ha/year of rice, removes from the soil nearly 660 kg/ha of N, P and K and quantities of several other micro-nutrients, such as zinc, iron, copper and manganese. Farmers only replace N, P and K. Research shows that every application of 100 kg/ha of NPK (at a ration of 50:25:25) leads to the depletion of zinc (by 629 gm/ha), copper (433 gm/ha), iron (4,180 gm/ha) and manganese (4,185 gm/ha). The greater the use of NPK, the higher the depletion rate of some micro-nutrients, which are not replaced by farmers. It has been estimated that the cost of restoring zinc alone in India would cost as much as $1,400 million.

Another technical issue to be considered is whether the future technology will find a way to grow crops without using soil. Such technologies would greatly reduce the impact of irreversible fertility loss in existing soils.

Causes of Soil Erosion

Soil erosion and the deposition of sediments in water reservoirs are not solely the result of farmers' actions upstream. Erosion is also a natural geologic process that will continue, regardless of human actions. A basic issue is how to distinguish natural erosion from man-made erosion[1]. The same problem arises when determining the causes of flood damages downstream. In this case, snow melting as well as high rainfall rates in the valley need to be taken into account. Otherwise, the economist may overestimate projects benefits.

Availability of Data

The nature of the soil erosion process is such that a comprehensive understanding of the phenomenon and its effects requires inputs from many different professions. The data needed to carry out technical as well as economic and social analysis are frequently lacking. Research and data collection efforts are weak and governments have allocated few financial resources to improve the data base on erosion-related matters.

Since many factors affect erosion rates, meaningful analysis requires both cross-sectional and time series data. Where the variability of soil erosion rates is being analyzed, cross-sectional data are useful to capture the differences due to different farming practices and different topographic and geographic characteristics.

In situations characterized by dynamic intertemporal changes in soil and in sedimentation over time, analysts also need time-series data. These data are particularly relevant in projecting the erosion rates that will occur "with" and "without" a soil erosion control program.

Surveys carried out to assess the soil status of any given land area or to establish a monitoring and evaluation unit should be used to collect both cross-sectional and time-series data. A review of a large sample of survey questionnaires designed to diagnose soil erosion problems showed that most of the questionnaires were designed to maximize cross-sectional variability. As a result, the data have proven to be of little use.

Practical Issues

Several practical issues arise in appraising soil conservation projects. Some of those issues are outlined below.

Cross-Disciplinary Measures

Given the multidisciplinary nature of appraisal, it is critical that the appraisal team develop a common focus. The economic staff must develop appraisal methods which utilize the information that the technical staff is able to generate, or conversely, the technical staff must develop technical measures which are useful to the economists.

The key to the development of such measures is a comprehensive understanding of the relationship between economic and physical measures.

Scope of the Effort

As obvious as it may sound, it is very important to define the scope of the task. One useful boundary may be to specify whose welfare is meaningfully influenced by the erosion process. The entire team, including the economist, needs to have a basic understanding of the operation of the physical system on which the project will be imposed. The logical way to handle such a complex risk is to develop a framework which relates physical processes to the welfare of economic agents.

It is equally important to define which are the most critical aspects of the task, to select those items that will make a real difference to the appraisal. Only parameters that have a potential influence on the project evaluation should be considered.

For example, the particle size distribution of sediment in runoff may be an interesting scientific question, but if there is no relationship between the economic costs and benefits relevant to project evaluation and particle size distribution, it cannot be considered a critical parameter.

On the other hand, if a major portion of the benefits of the project can be related to improved water quality and specifically turbidity, such a parameter may be an important variable.

Developing Practical Measures

An economic appraisal is used to predict the economic results of a soil erosion project. Unless important variables are measured and related to a specific notion of economic welfare, the evaluation process will be useless. The relative complexity of measurement will depend on the particular circumstances of the project. Simple and descriptive frameworks may do the job in certain areas, while meaningful results will only be obtained with more complex interactive models in other cases.

These decisions, related to which analytical framework to use, have important implications for defining the data-gathering process. Data gathering is complex and expensive. To justify an expensive data gathering procedure and the operation of advanced computer models, simpler methods must be shown to be inadequate.

Worksheets showing cause-effect relationships may be helpful in determining the most important areas of future analysis. Computer software technology has advanced tremendously during the last few years. Computer modelling, therefore, is no longer a thing of the future but is now a practical tool. Farm models, hydrological models and models to simulate farmers' behaviour are now inexpensive and easy to operate even in the field.

Dealing with Unprofitable Projects

Measures of project profitability (for example, IRR, NPV) are only one indication of project worth. Because of the nature of erosion control programs (for example, irreversibility, intergenerational equity), one may be required to accept an "unprofitable" project in NPV terms because of other economic and social considerations. Further, a project may be unprofitable in a narrow sense -- due to inaccurate cost and price projections in an environment characterized by a great deal of uncertainty, or to discount rates that are too high -- and thus fail to reflect society's "true" time preferences, or many other factors.

Therefore, in order to make an adequate analysis of "unprofitable" projects, one should not only examine immediate economic returns but also should consider benefits and costs as they accrue over time. Sensitivity analysis is of principal importance here. The main issues are: What information should be presented to decision makers regarding possible effects above and beyond the specific project's NPV? What demand rates, price projections and quantity projections should be used and how should they varied?

Notes

1. See Jakobsson and Dragun (1991).

13

Organizational and Institutional Issues

This chapter focuses on organizational and institutional issues. Two institutional issues of particular analytical importance, compensation schemes and intergenerational equity issues, are discussed in a later chapter.

Organizational Issues

Very few developing countries have the organizational foundation needed to deal with the many facets of the soil erosion problem. Because of the multidisciplinary and multisectoral nature of the problem, a unified organizational framework is needed. At present however, the frameworks in developing countries are fragmented and government departments often do not know about nor co-ordinate their soil conservation activities with those of other departments.

The Role of Government

Government intervention in soil conservation activities has often been justified on two grounds. One is that the market for goods and services fails to provide signals to producers and consumers on the necessity of narrowing the gap between social and private values. External effects -- where actions by one economic agent affect the level of economic activity of other agents in the economy -- are ignored. The other is that soil conservation programs are generally long-term in nature and the government is required to intervene to ensure that the rights of future generations are not violated because the market fails to make resource allocations that are equitable across generations. Within this context, five issues deserve specific attention, resolving conflicts at the policy level, unification of approaches to carry out programs, integration of resource management planning and overall development planning, budgetary allocations and data bank needs.

Why public intervention is needed at the national level and at the sector or district level. Experience shows that the success of soil conservation programs is closely associated with good program administration. In a range of developing countries the main constraint may be seen as a lack of trained personnel to carry out soil conservation programs at the local level. The tradeoffs needed to resolve conflicts between private and social interests are at the heart of the problems. The main issue, therefore, is how the government can act effectively without jeopardizing private sector initiatives.

The public sector in developing countries is often fragmented. A characteristic of soil conservation programs is that too many separate agencies are involved with little coordination by the central government. A unified approach is required because soil conservation programs must be multidisciplinary and multi-regional. Generally, the natural sciences aspects of soil conservation programs are usually dealt with by ministries of agriculture and forestry, while the engineering and mechanical aspects of soil conservation programs often are the preserve of ministries of public works, energy, or irrigation. This fragmentation is often seen when a hydroelectric project is implemented. In most cases it is the ministry of energy or public works that oversees the project, with major emphasis on its engineering aspects. Very little effort is made at the same time to control erosion through the establishment of a comprehensive watershed protection program.

If investment and policy planning are to take into account the role of natural resources, land use planning must be integrated with the overall development planning process. At present, different ministries tend to act in accordance with their own interests and their attempts to participate in comprehensive policies from an isolated position invariably decrease their effectiveness in dealing with soil erosion. Several developing countries do have natural resources laws or codes (Columbia, for example), but these are not taken into account in the overall development process.

Administrative coordination is extremely important if soil conservation policy is to be effective. Administrative coordination is not easy to achieve, since natural resource planning represents only one portion of development planning and many factors complicate the planning process. Many proposals have been made on how governments can achieve more effective policies. One proposal is that developing countries should create soil conservation authorities. These authorities would be similar to the regional authorities that are responsible for comprehensive development of river basins. One important feature of such authorities is that their structure unifies the approach to a given natural resource.

Planning often requires definition of a "unit of account" which defines the scope of the plan. One often sees the use of "national," "regional," "state," local," and "farm" planning. The selection of a proper unit of account is particularly important in the management of natural resources and in the assessment of policy effectiveness. Watersheds have now become an important unit of account in the planning, organization and development of soil conservation programs and projects. This geographical classification has conceptual unity, particularly from the point of view of planning hydrological changes. The importance of watersheds has become particularly evident because of the development of hydroelectric power facilities. From an economic and institutional perspective, two important things should be understood, (1) The people and activities in the upper portion of a watershed are different from those downstream, (2) the major emphasis in development planning has been promoting production rather than on guiding the price signals associated with those changes to optimize the use of scarce natural resources (for example, for energy, for forest products, or for food). From an ecological viewpoint, a watershed as a unit of account may prove to be much more effective than other units of account (See Ciriacy-Wantrup, 1959).

The size of unit of account is an important factor in increasing the effectiveness of soil conservation planning and implementation. On the one hand, it may be important to consider the creation of wider planning boundaries, since the soil conservation problem may go beyond the political boundaries of one state or province. On the other hand, there may be a need for smaller units to improve program implementation, since programs covering large geographical areas may require more local resources than are available. These decisions need to be carefully assessed.

An important characteristic of soil conservation programs is their annual need for budgetary allocations. Governments are often not well organized to allocate funds between capital and recurrent expenditures at the national or local levels. This problem is compounded by the fact that, more often than not, there are no local level organizations to implement soil conservation programs. In contrast, several developed countries have created local "soil conservation districts" with clearly defined functions and budgetary allocation procedures to carry out such programs.

An important role of government is to maintain a data bank to provide meaningful information for project/program planning and implementation. A data bank is particularly important for soil surveys, land use classification and hydrological and sedimentation studies. Given the nature of the erosion problem, one should not expect the private sector to carry out this activity.

Agricultural Research

Soil erosion is in great need of applied and goal-oriented research. There are many soil erosion studies, particularly technical ones, where the results have not been tested. Until they are, useful findings cannot be turned into activities at the farm level. National and international agricultural research is often "crop-oriented." Problems such as soil erosion are seldom studied in an integrated manner (that is, taking into account both upstream and downstream effects of erosion).

Research should focus not only on the technical aspects of soil erosion but also on its socioeconomic and institutional aspects. Examples of socioeconomic research are farmers' behaviour in relation to the adoption of soil conservation practices, the role of demographic factors, changes in consumption patterns and the effects of farmers' organizations and attitudes. There has been little or no research on how prices, taxes and subsidies affect farmers' adoption of soil conservation practices. Few research studies have focused on the potential impacts of changes in tenure, property rights and tenure uncertainty as they relate to soil conservation (or depletion) practices.

An extensive review of the literature shows that most of the research that has been carried out is "micro" oriented. Many scientists have devoted time and resources to assessing the many factors that influence soil erosion rates, such as rainfall, slope, soil chemistry and the like. Most of these studies have been designed to test for one of these factors, holding everything else constant. Because of their small scope, such studies have seldom produced results applicable on a global basis. Since much of the world's erosion problem is occurring in rainfed agricultural areas (for example, in Africa), experts state that much more research needs to be carried out on the technical as well as the economic viability of rainfed farming systems.

Research into soil erosion and related problems is often costly and requires long-term budgetary allocations. Such an apparently simple matter as monitoring sedimentation rates in major rivers or water reservoirs, for example, is expensive. Such monitoring requires sonar and other types of sophisticated equipment and skilled personnel. Project analysis depends on long-term time series data. Thus, several years of observation of soil erosion and related phenomena are required before the data is useful for investment planning.

Extension, Education and Training

The creation of public awareness about the nature and magnitude of the soil erosion problem is a necessary condition for project or program success. Extension services, education, training are the key ingredients in creating this awareness.

The achievement of specific output targets is the main objective of the large majority of extension services in developing countries. Most extension systems, in other words, are essentially production-oriented, responding to an immediate need to produce more output per unit of land. Very few extension schemes, if any, are conservation-oriented. Thus, in areas where there are significant tradeoffs between production and conservation, few efforts are made against soil erosion. Changes in the extension service approach are needed. These should be accompanied by the provision of adequate funds and staff and a clear integration of extension activities at the regional and local level.

The "unit of account" used by most extension services is the farm. However, as pointed out earlier, addressing the issues related to the soil erosion problem requires changing the unit of account to some ecological unit, such as the watershed. Because of the external effects of erosion, the individual farm is not necessarily the right unit of account for an optimal program. The proper definition of the unit of account is more important in areas characterized by *common property* in land resources.

Policy makers, administrators, farm managers and others who influence public opinion need a better awareness of environmental problems. This educational process has to be implemented at all levels and should provide an integrated view of environmental phenomena. Education and training have focused mostly on short-term problems and needs and have usually become very specialized. The frequent outcome is piecemeal actions that are extremely ineffective in managing natural resources.

Monitoring and Evaluation

Monitoring and evaluation activities are important instruments of project and program management. Few developing countries have monitoring and evaluation units that are capable of providing management with data and performance indicators on the progress of soil conservation programs. Field experience in a number of countries shows that monitoring and evaluation activities are fragmented. Consequently, government agencies do not have the necessary data to make decisions or to assess the extent to which soil conservation programs are achieving specified objectives.

Data on sedimentation rates -- when they are collected at all -- are often collected by ministries of public works or energy, since sedimentation affects the economic life of hydroelectric dams. These dams are seldom integrated with data collected by the ministries of agriculture and forestry which focus on the intra- or inter-farm effects of soil erosion, even though the latter ministries are often in charge of the implementation of soil conservation programs at the farm or watershed level. The main issue therefore is to establish an overall organizational framework for the collection, processing and use of data now being accumulated by different departments of government. Sufficient public funds to carry out these activities in a sustained manner are essential.

Finally, it is important to mention that an important part of an erosion control project is the formulation and implementation of monitoring and evaluation units. Data collection often requires field surveys. Most surveys are designed to obtain information about on-farm production, with very little information being collected about off-farm effects. Therefore, an effort should be made to design surveys that will provide the necessary data to monitor and evaluate both the on-farm and downstream effects of erosion.

Farmers' Organizations and Participation

Since it is the farmer who must adopt recommended soil conservation practices, farmers clearly need to be encouraged to participate in soil conservation planning and decision making. The extent of farmers' participation is something that needs to be defined, bearing in mind country and local social conditions.

Lack of farmer participation in soil conservation planning inevitably means that farmers see recommended practices as being imposed from the outside. As a result, farmers very often abandon these practices soon after the program begins. Economic incentives are also a key element to success. If conservation practices are perceived by farmers to be unprofitable, these practices will not be maintained. Thus, incentives thus are especially important, because the effectiveness of many soil conservation practices depends on sustained implementation over long time periods. This applies in particular to low-cost physical infrastructures (for example, gully control structures, bench terraces) where maintenance requires large amounts of labor by farmers and their families.

Soil Conservation Planning and Development Planning

Soil conservation planning must be integrated with national development planning. The environmental sustainability of soil resources and of land in general, should be a main subject in any development plan. Much agricultural planning concerned with physical inputs and outputs aims at output maximization. Most plans focus on the major determinants of different forms of capital and labor. Very few plans integrate management of these inputs with the preservation of natural resources, in this case with land. In both microeconomic and regional planning, much more time is devoted to assessing what "goes onto the land" (for example, tractors, seeds) than assessing the impacts of these plans on the existing natural resource base available to rural people.

On the other hand, general policy analysis is often purely macroeconomic, dealing with financing, consumption, production and trade. Natural resources are generally ignored in the planning process and a real awareness of their existence only comes about when output targets are not achieved or where irreversible damages occur. This is very shortsighted and the emphasis on this type of planning framework is responsible for the tremendous degradation of natural resources that is taking place today.

Coordination Among International Donors

Coordination of international aid is also needed. In some instances, international finance agencies pursuing the same global objectives may end up counterbalancing each other or even producing negative effects on the rate at which erosion is controlled. The key topics are, (1) compensation schemes, (2) incentive programs and (3) intersectoral coordination.

Although the device of compensating farmers for adopting new practices may be a necessary condition of success for a particular soil conservation program, incentive schemes across regions or countries should be coordinated. It is not uncommon to find donor agencies concentrating their efforts on different project areas. Within these areas, agencies establish their own compensation mechanisms. Differences in the amounts or type of compensation however, can result in income changes that cause shifts in migration patterns -- movements of people from areas of lower compensation to areas of higher compensation. Shortages of labor may then occur in outmigration areas and labor-intensive conservation practices will not be carried out. In immigration areas, on the other hand, increases in population density may increase the pressure on lands which conservation programs were supposed to protect. Such uncoordinated effects may defeat the original purpose of conservation programs.

The same need for coordination applies to incentives. Lack of proper coordination of market and nonmarket incentives may change the local economic environment to such an extent that conservation practices are not carried out, or are counterbalanced by programs with other goals. One agency may be implementing a price support system that increases the production of soil-eroding crops while another agency is trying to discourage the cultivation of such crops through other mechanisms.

Several types of conservation schemes require intersectoral coordination. One donor agency may be funding the construction of a multi-purpose water reservoir while another is funding increased exploitation of the watershed upstream. Under these circumstances, it will be very difficult to control the rate at which sedimentation takes place.

Ancillary Activities

Often, soil conservation programs do not take ancillary industries, such as those devoted to producing farm supplies and farm equipment, into account. The programs then become difficult to sustain. To carry out contour ploughing on a steep slope, for example, one needs a special type of mobile plough. If this type of plough is not available to farmers they will abandon contour ploughing, since implementation with their existing equipment requires too much time and energy.

A similar problem occurs when soil conservation programs focus on replacing soil nutrients. Such programs depend upon fertilizer supplies. If fertilizer is not available in adequate quantities, the whole purpose of the conservation program can be defeated.

Institutional Issues

As Ciriacy-Wantrup (1968) points out, there are two important types of institutions, social institutions and economic institutions, which are relevant in this class of problem setting. The latter most often operate through the market ("market institutions"), while the former do not ("nonmarket institutions"). Both kinds of institutions are decision systems in the context of soil conservation policy decisions. Examples of market institutions are prices, interest rates, income, taxes, subsidies and credit. Examples of nonmarket institutions are tenure and property rights.

These institutions comprise the overall incentive structure faced by farmers when making decisions about the conservation or depletion of natural resources. Since the role of market institutions is well-known, the remainder of this chapter focuses on issues relating to a few nonmarket institutions.

Typology of Effects and Institutions

Earlier we presented a typology of upstream and downstream soil erosion effects. This typology was developed because many analysts fail to identify all the major benefits and costs of soil conservation programs. In other words, the typology originated in the need to improve analytical frameworks used in the preparation and appraisal of these programs.

This typology is related to institutional arrangements for the successful implementation of soil erosion control programs. To deal with on-farm effects, for example, institutions designed to change the individual farmer's behaviour must be established. On the other hand, institutional arrangements designed to deal with interfarm effects should focus on farmers' organizations or other collective institutions. To deal with downstream effects, a regional or national institutional arrangement is needed to regulate the actions of various sectors in the economy.

Thus, the typology makes it possible to distinguish among alternative institutional arrangements and to assess their capacity to achieve development objectives.

Tenancy

The degree to which incentives to conserve soil are effective is greatly influenced by tenancy -- or land tenure. It is well-known fact that uncertainty of land tenure leads to soil depletion. Because tenant farmers are not certain as to their entitlements with regard to use of the land, they try to maximize short-term production gains and tend to disregard long-term investments. The same would apply to landlords who were uncertain about their long-term rights because of political uncertainty. High erosion rates have been found in areas where tenancy arrangements are not secure (that is, where short-term leasing of land is common). A typical situation in many developing countries is "minifundia," where landholdings are so small that it becomes uneconomical for farmers to adopt soil conservation practices. Demographic pressures, combined with laws or customs with regard to land inheritance, exercise tremendous influence on the way in which soil resources are used.

In some instances tenurial arrangements are not the most important determinant of proper land use. Subsidiary markets for financial capital play an important role to access new or under-utilized lands. These markets will develop if trading conditions and liquidity are allowed to develop. The development of financial markets is an important element often disregarded and an expansion into unused lands could in some cases greatly alleviate the need for more intensive land use in areas of very high demographic density.

Property Rights

Property, in its different forms, is a primary social institution. Different countries -- and farming systems within those countries -- have different systems of property rights.

Among the many types of property rights are *private rights*, *common rights* and *public rights*. Although there is general agreement about the definition of private rights, there is disagreement about the meaning of "common property." Many people describe common property as no one's property and "free access" to it is said to be its most important feature. This is a mistake. Without pretending to resolve this disagreement, we note here that where there is common property in land resources, the property rights are vested in given individuals -- who might be considered co-equal owners (See Jurgensmeyer and Wadley, 1974 and Ciriacy-Wantrup and Bishop, 1975). In a system of common property, as in any other system of rights, there are specific rules and regulations with regard to access and to appropriation of potential rent from the resource -- as well as decision processes to modify or change such rules.

Common property rights exist in many areas of the world, particularly in those regions where common grazing of livestock is practiced. The numbers of cattle owned or managed by each herdperson will greatly determine the productivity of particular soil resources over time. In many parts of Africa the number of animals an individual possesses is associated with social status. In such cases, there is an incentive to maximize the size of one's herd, resulting in the overgrazing of existing pasture lands. Thus, it is important to recognize and analyze the rules governing the use of pastures.

One of the biological methods used to control erosion in many regions of developing countries is that of tree planting. In many countries, such as India and Nepal, successful implementation of such programs depends on the nature of property rights. Forestry programs have been designed for public lands, for communal or village lands and for privately owned lands. Property rights not only determine who owns the land (and the trees, in some cases) but also establishes a mechanism for the distribution of the proceeds (or rent) derived from land use. The distribution mechanism, whether in cash or in kind, plays a more important role in situations characterized by public or communal property than where private individual rights are involved. On privately owned land there is a one-to-one correspondence between who owns the rights and who controls the proceeds, the private owner. Although evaluation of private forestry schemes has been fairly random, the empirical evidence suggests that these schemes have been quite successful. On commonly owned land, in contrast, the lack of clearly established mechanisms for the distribution of proceeds has made tree conservation more difficult.

Common property has often been cited as a source of problems -- *the tragedy of the commons*. It is often said that common ownership is to blame for exhaustion of resources. Common grazing land, tubewell irrigation and exploitation of fisheries resources are often cited as examples of how common use can deplete resources and lead to overcapitalisation of exploitation. The natural recommendation that often follows is to change common property rights to private property rights and let the "invisible hand" allocate the resources. It is our contention that:

1. There is a misconception as to what common property really is.
2. Common property schemes are not *per se* the cause of resource depletion.
3. Common property is an important framework for policy formulation and solutions.

First, "common property" (*res-communes*) is often confused with "nobody's property" (*res-nullius*). Common property means coequal ownership and requires the exclusion of those who are not co-owners of the resource (in this case, soil or more generally, land). The rules regarding the use of common property are well-established either by custom and tradition or by law. Second, it is not common property as an institution that is the major cause of resource depletion. There are many examples of resources that have been held in common for years, yet no depletion has taken place: "England and Wales still contain 1.5 million acres of commons, the bulk of which is being grazed much as it has been for centuries and they are still highly productive"(See Ciriacy-Wantrup and Bishop, 1975). What is often responsible for resource exhaustion is exogenous shock, the introduction of profit-oriented markets, changes in production technology (for example, tubewell development), or other exogenous factors (for example, monetization) that have created instability and resource exhaustion. Finally, many countries, including the European countries and the United States, have used the concept of *res-communes* as a basic framework for resource management policies. This should often be done in developing countries. The correlative rights doctrine for groundwater development in California and the establishment of fisheries quota systems are two good examples of how the concept of common property can be used to develop a policy framework.

In sum, the question of whether any system of property rights will have a good effect on soil conservation decisions depends on time and stability. When rights are unstable, uncertainty ensues. In such a situation, farmers will prefer to maximize short-term gains of which they are sure rather than try to maximize society's long-term objectives, which they cannot identify as similar to their own. The devastation of the "commons" has occurred in many countries not because of the intrinsic characteristics of common property but because of the imposition of what local inhabitants felt was a "foreign" set of rules.

Fundamentally, the land management "problem" may be conceived in a commons situation where a range of individuals in a region may own their land and have a common interest in the negative [or positive] effects of their combined use of the land.

Credit

Rural financial savings are often in short supply, mainly because farmers' incomes are low. In carrying out soil conservation practices which will come to maturity very far in the future, intertemporal liquidity at the farm level is an important factor. Credit is one of the most important sources of liquidity. The potential advantages for erosion control of expanding liquidity in the system will depend on timely allocation of credit proceeds and sensible repayment procedures.

Credit policy should take at least two major factors into account. First, it is important to distinguish between short-term seasonal credit for input supplies (such as fertiliser) and longer term credit for conservation practices that take the form of land investments (for example, gully control structures). Very often, this distinction is not made and repayment procedures do not match the intertemporal creation of liquidity resulting from the soil conservation practices. Second, it is necessary to understand that the level of interest rates (that is, the cost of capital) will play a very important role. If interest rates are too low, farmers may use more capital-intensive equipment, which intensify the rate of soil erosion. But if rates are too high, the value of liquidity in the short term will rise, thus giving farmers an incentive to increase production. This too will put pressure on eroded soils, defeating the purpose of the conservation program. Such tradeoffs therefore need to be clearly assessed.

Legislation

The enactment of laws is a way to formalize rules, regulations and incentives that will influence farmers' soil conservation decisions. The enactment of comprehensive laws governing the allocation and use of natural resources in general and of soil resources in particular, is a difficult and costly process.

A comprehensive look at the laws in developing countries indicates that resource conservation legislation is often absent and when in place, is often incomplete. This is because developing countries do not have well defined principles for managing their natural resources. Apart from a lack of accompanying codes and regulations on implementation, existing laws on soil conservation usually do not have a clear statement of purpose, do not identify the major sources of erosion, do not establish priority areas, do not provide mechanisms to determine violations of major objectives, do not define government financial or administrative responsibilities, do not define intra- or interagency coordination and do not provide a framework for implementation or enforcement.

In many cases, the laws are based on norms which are inconsistent with the ways in which traditional groups informally handle production, use and distribution of natural resources in rural areas. The lack of correspondence between formal and informal rights creates conflicts during program implementation.

Laws governing the use and allocation of natural resources in many countries are often "sanction-oriented," establishing only the procedures for punishing those who violate certain principles. This is a very primitive form of legislation. In contrast, a few laws (for example, in Argentina) have been drafted only as "incentive oriented" frameworks -- that is, giving farmers incentives to adopt soil conservation practices. Problems arise in either case, since the laws should contain both incentives and sanctions.

Even if the legislative framework is adequate, there are several reasons why it may become ineffective or difficult to enforce. One of these is that individual offenses are often very small in terms of monetary value. For example, if a law enforcement officer finds an individual illegally cutting a few tree branches for fuelwood, the state's claim against the individual may be less than the litigation costs. This low ratio of the costs of damage to the costs of enforcement often results in no action by the state. Therefore, laws have to provide for an enforcement mechanism. In some soil situations, the local community could play an important role in enforcing the laws, for example, when a community group is formally responsible for the repayment of loans or the payment of water use fees and must collect the money by informal means from individual members. It is clear that the proportion of sedimentation in a reservoir that comes from, or could be attributed to, any individual farmer's actions is so small that the cost of prosecution would again be too high to carry out. Here again, it may become important to write laws that apply to groups of farmers or some other collective group rather than to individual farmers.

These aspects of resource conservation legislation are an important subject for future research.

Tenure and Soil Conservation

Security of tenure, including the whole range of legal and institutional factors that condition access to and control of land by individuals or groups, is widely thought to be a key determinant of landholders' willingness to adopt soil conservation measurers. Soil conservation includes a wide range of investments or methods intended to hold soil in place, thereby protecting the long-term productivity of the land. Thus, it is often assumed that if farmers face a great deal of uncertainty with regard to their continued possession of a piece of land, they will be less interested in adopting soil conservation practices. This will be especially true in the case of long-term, capital-intensive investments, such as the planting of certain tree varieties or the construction of terraces. Unfortunately, very little empirical data are available to either confirm or refute a connection between insecurity of tenure and adoption of soil conservation practices. However, there are numerous pieces of anecdotal evidence that suggest ways of taking tenure issues into account in preparing soil conservation programs.

Anthropological studies in Kenya (Brokensha and Riley 1978), for example, have found that land use patterns are intimately tied to forms of land ownership. When formal title was awarded to landowners in the Mbere region, the customary rights of nonowners changed drastically. Prior to awarding a formal title, for example, the right to collect firewood on any land was essentially unlimited. Following the assignment of title however, the rights to collect firewood or wood for building and to collect wild plants, became strictly controlled. In Thailand, a principal rationale for a land reform program and entitlement project was to give security of tenure to promote adoption of soil conservation practices.

The connection between tenure and soil conservation was also examined in Haiti (Murray 1979). Although no clear evidence of a link was found, the study illustrates the complexity of the issue. Here, it appears that although formal deed are widely held, there is widespread fear of expropriation stemming from long experience with hostile public and external groups. Despite this fear and insecurity, a thriving market for land exists. This suggests that while the purchase of land may be risky, farmers perceive sufficient reward from ownership to induce them to invest large sums of money. Murray found that insecure tenure and a need for current income combined to make only those conservation measures with short payback periods acceptable.

In cases where tenants rent the land, there are other disincentives to soil conservation over and above time and uncertainty. For example, implicit in many rental agreements is the notion that tenants will be given first rights to purchase if the owner decides to sell the land. However, if soil conservation investments have been made, the price of the land will be greater. Thus, some tenure arrangements can result in a reluctance to conserve as a hedge against the opportunity to eventually purchase land.

Another example of where the decision to invest in soil conservation measures is linked not so much with land tenure as with ownership of the land itself is illustrated by a World Bank-financed project in India. In the Uttar Pradesh Social Forestry project, farmers' reluctant to plant trees has been attributed to an ongoing project of land consolidation. While farmers with clear title to land are willing to plant seedlings, those who are concerned that their land holdings may be affected by consolidation are not. This stems from concern that should their land be redistributed they will lose the rights to the proceeds from the trees they have already planted.

These experiences suggest that land tenure does indeed play an important role in determining the acceptability of soil conservation measures to landholders, but that the nature of the land tenure role is not obvious and is subject to the specific nature of the terms on which land is occupied. A key issue is that of expectations about future rights of access. Uncertainty about property rights, in regard to *who* will be involved in determining access and control and in regard to *when* the uncertainty itself will be resolved, all seem correlated with reluctance to invest in soil conservation.

Another effect of tenure insecurity is a reduction of the time horizon of farmers. In a sense, insecurity raises the tenant's rate of time preference, leading him to discount faster and at a much higher rate the potential gains from long term investments such as trees or terraces.

The issue of tenure has several implications for project planning and analysis. The overall issue is simply the need to realize that projects are carried out in a complex environment, are affected by a large number of factors, one of which is land tenure and that efforts must be made to anticipate these effects. More specifically, it seems reasonable to believe that some land tenure arrangements are more conducive to achieving efficient levels of soil conservation investment than others and that legal and political reforms to promote more efficient forms of land management are justified.

In common property lands, security with regard to distribution f the proceeds is an important element. In a communal forestry project in India, contracts between villages and the Forestry Department were prepared in order to convince villagers that their rights to trees planted using their labor on these communal (or public) lands would be preserved. These contracts specify rights and obligations with respect to the provision of inputs to the project and the distribution of outputs.

While the fundamental source of insecurity in land tenure may not be easily removed and may only be alleviated by using controls and enacting proper legislation, certain options are available to reduce insecurity's effect on adoption of soil conservation measures. For example, given a choice between biological interventions with immediate effectiveness (such as mulching) versus physical interventions with long payback periods (such as terracing), explicitly consideration of tenure insecurity may be critical to effective project design.

Finally, it is clear that tenure insecurity may drive a wedge between private and social costs and benefits. It may, for example, result in private rates of time preference being higher than social rates, or in the returns from soil conservation being inefficiently distributed. When this is the case, analysis of projects from a range of accounting perspectives is required and can assist in the design of efficient incentive and compensation schemes.

Strengthening our understanding of the link between land tenure and incentives for conservation is a clear research need. Empirical research in the United States (for example, by Lee, 1980) has generally been unable to establish a clear link between form of tenancy and conservation practices. Simpson and Young (1973) however, in their analysis of open access grazing land, have shown that the profitability of conservation measures is related to systems of ownership.

Ervin (1982) did some empirical research to assess in quantitative terms the differences in erosion rates between owner-operated and rented cropland in Missouri. One important conclusion of his study was that, on average, the erosion rates on owner-operated farms were much less than on rented farms. In cropland and pasture the difference was 41 percent, while for cropland only the difference was 29 percent. This has very important implications, since Ervin estimated that nearly 40 percent of US cropland is now rented or will be rented in the near future. However, it was not totally clear if such a very large difference in erosion rates was due to the nature of rental arrangements or the personal characteristics of farmers.

The above mentioned results have specific implications with regard to the conservation programs, local or national, that governments decide to implement. In cases where such tenure problems exist, it may be advisable to promote low-cost programs, such as minimum tillage and contour farming. In addition, where compensation or cost-sharing schemes are formulated, they should not be restricted to owner-operated farms.

The New South Wales Legislation: Carrots and Sticks[1]

Here we present several extracts from the Soil Conservation Act, 1938, No. 10, State of New South Wales, Australia. These may be helpful to other legislative bodies considering enactment of laws, rules and regulations.

One aspect of the Act is the fact that the Soil Conservation Service (SCS) is an autonomous institution, responding directly to the authority of the Minister. The Act states that the Governor may from time to time appoint a Commissioner of the SCS who shall, subject to the control and direction of the Minister, exercise and discharge the powers, authorities, duties and functions conferred and imposed upon him.

With the view to conserving the soil resources of the State and protecting catchment areas and proclaimed works, the Commissioner may, of his own notion and shall, if so directed by the Minister, conduct experimental and research work in connection with soil conservation and erosion mitigation. He shall also conduct such investigations relative to any aspect of the erosion, publish such information as he may think necessary of the results of any experimental or research work, conduct demonstrations of methods of soil conservation in erosion mitigation, and carry into effect projects in any part of the State.

To carry out some of the necessary works, the SCS, via the Minister's decision, will also manage financial resources. The Act states that the Minister may, out of the monies provided by Parliament, make an advance upon such security and at such a rate of interest and subject to such covenants, conditions and provisions as he may think fit,[2] to any *owner or occupier* [emphasis added] of land proposed to be dealt with. The emphasis is added since many pieces of legislation do not focus on the owners of the land. In many developing countries, monies are not channelled to those who do not own the land.

The Act, in addition to recognizing the implementation of work dealing with erosion in a strict sense, also considers as part of the necessary actions, works in, a) water conservation, irrigation, water supply or drainage, or b) the prevention of inundation of land and overflow of water thereon, or c) changing the course of a river. Within this context, the Act empowers the Minister to let or hire any machinery, plant or equipment or provide labor, material or services to any person for the purpose of the carrying out of works, subject to payment of such charges or costs as may be agreed upon.[3]

An extremely important concept within this legislation is the concept of "area of erosion hazard". The Act states that where the Minister is of the opinion that any tract of land is subject to erosion or is liable or likely to become liable to erosion and that such tract of land should be notified as an area of erosion hazard may, by notification published in the Gazette and in a newspaper circulating in the locality in which such tract of land is situated, give notice of a proposal that such tract of land should be notified as an area of erosion hazard.

Where the owner, occupier, or mortgagee of any land within the area of erosion hazard has neglected or refused to enter into any agreement under this Act, or has failed to carry out any covenant, condition or provision, the Minister may authorize the Commissioner to carry out remedial work on the land, to adopt "new" practices of land utilization or land management, or to limit the livestock which he shall carry on the land.

Cost recovery of such works is also an important element in this Act, the Minister may recover the costs incurred from the owner, occupier or mortgagee of such land in any court competent jurisdiction as a debt due to the Crown and until such repayment, such cost shall be a charge on the land.

Finally, it may be worthwhile to note a few actions relating to water conservation. The Act states that the Governor may declare, a) any work or proposed work for the storage, regulation or conservation of water, or b) any work or proposed work for or in connection with any one or more of the following, (1) preventing or mitigating the inundation of land or the overflow of water, (2) defining or changing the course of a stream or river, (3) maintaining or improving the banks or foreshores of a stream, river or harbor, or (4) deepening or maintaining the depth of a stream or river.

Notes

1. The material here is paraphrased from Australia, Soil Conservation Act, 1983, No.10. Reprinted under Reprints Act, 1972 [Reprinted as at 25th February, 1981] New South Wales.

2. The SCS Annual Report 1980-81 states that to assist landholders unable to finance soil conservation programs, the Service has a scheme for long-term loans which is administered through the State Bank. During 1980-81, 406 advance applications were approved and an amount of $3,728,146 (Australian) was advanced.

3. The SCS's Annual Report for 1980-81 states that it operates a fleet of 102 crawler tractors and ancillary plants. The plant with operators is hired to land users to assist them in implementing the economic control programs. The scheme operates on a commercial basis.

14

Macroeconomic Policy Issues

Managing Natural Resources

Arguments for resource conservation and stewardship are often phrased in ethical terms. The argument is usually along the following lines. Agricultural sector outputs are extremely important to society as a whole, since they satisfy basic needs for food and nutrition. Social pressure to maintain the productivity of the agricultural sector over time is logical and obviously soil is an important agricultural input. Inadequate management of natural resources can result in irreversible losses that will have negative economic and social effects.

Resource conservation raises several broad policy issues, how to prolong the useful life of the natural resource base, how to avoid irreversible damages to it and how to maintain options for future generations. All three apply to soil conservation programs. In addition, it is important to note that the design of soil conservation projects may change sedimentation rates in ways that harm navigation channels or eliminate potential dam sites.

Evaluation should take into account the social value of maintaining future options. Unfortunately, it is difficult to specify such social value in quantitative terms. In many countries, furthermore, this issue is not well understood.

Establishment of Priorities and Standards

To achieve optimum conservation of soil resources, planners need to identify the main soil conservation needs. Since financial and human resources are limited, priorities will have to be set. In other words, it will be necessary to identify the most important soil conservation activities at the national, regional and local levels. Because soil erosion and sediment control policies often have important distributional impacts, standards have to be set with a clear understanding as to who benefits and who loses. Priorities which do not take into account the potential incidence to politically powerful groups in society, tend to be ignored.

Standards are an essential ingredient of any comprehensive plan for soil conservation. The standards should designate which types of soil should be preserved or rehabilitated. It is unrealistic to establish standards calling for the total elimination of erosion if the achievement of such standards is financially not feasible. Standards should also have some relation to economic benefits and costs. To target a soil conservation program -- to achieve a particular erosion rate, say 5 tonnes/ha/year -- could be meaningless unless the target makes sense relative to the available resources.

In many cases a lack of comprehensive economic policies and programs may result in an irreversible loss of important natural assets. These irreversible losses should be avoided whenever possible. This means that it will be necessary to make a careful assessment of tradeoffs in the allocation of government funds over time. Some governments may be required to spend more now in order to save later, despite a context of political instability and strong short-term needs. Because many governments are concerned about remaining in power, they often adopt policies designed to maximize immediate political dividends. But resource conservation and management can only be successful when it is understood that the goals are long-range matters.

Resource Management

The management of natural resources, including soil, requires new approaches. Land should be regarded as an asset, not simply a factor of production and it should be understood that irreversible losses of soil will narrow the options of developing countries. At a minimum, soil losses will translate into higher food production costs or larger payments for imported food.

In most countries, conservation policies are oriented more toward control of environmental degradation than toward prevention. Although control of soil erosion should be a very important goal, the prevention of future damage (often at much lower social cost) should become the centrepiece of policy making and planning. As explained earlier, the SMSC principle (avoidance of irreversible damage as the first priority) provides the framework for a sound resource management policy.

Existing natural resources, including soil, are often ignored in the formulation of macroeconomic policies. The tradeoffs between alternative macroeconomic policies and their effects on erosion rates need to be assessed. As in other countries, macroeconomic policies in the developing countries must be concerned with food production. In the developing countries the need for food is usually met through the production of grains, including grains which are relatively soil depleting (other things being equal). In many areas however, the demand for food is so great that rural people will continue to grow food on soils which are practically exhausted.

Export policies are seldom designed to take into consideration the potentially negative impacts of declining soil productivity. The need for foreign exchange results in export incentive programs tilted toward crops or farming systems (for example, mechanisation) that deplete soils over the long term. But the theory of comparative advantage rests on the idea that it is a country's relative abundance of natural resources that give it a trade advantage. Thus, agricultural export policies should be assessed in the context of proper management of the natural resource base.

As explained before, the dimension of time is of fundamental importance in policy decisions. An economy that achieves important short-term gains may do so at the expense of long-term social costs. An example of such a policy dilemma can be found in the forestry sector. Long-term real increases in wood prices may drive a country to increase its forest products exports at a rate that far exceeds the rate at which used forest lands are reclaimed. This will lead to greater soil erosion and may mean potential losses of other kinds of exports.

Macroeconomic Tradeoffs

International Dimensions of Management

Almost always, the "unit of account" for macroeconomic actions is the country itself. The adoption of this unit of account is natural, since it makes it possible to deal with many important factors. Where natural resources are concerned however, macroeconomic policies will only be successful where the "unit of account" exceeds national boundaries.

A case in point is the management of sediments in international river basins. To assume consistent policies, the basic unit of account should be the entire basin. It is very difficult for a country to manage the volume or type of sediments in a basin when the upper reaches of the basin are geographically located in another country. In other words, efficiency of sedimentation management policies will be reduced unless all of the countries that contain a part of the basin coordinate their efforts.

Conservation as an Income Policy

Resource management policies in general and soil conservation policies in particular should not be established independently of other economic policies -- this is particularly important in relation to income policies. The lower the per capita income in a country, the higher the social time preferences of people (the unwillingness to postpone consumption). To achieve the desired level of consumption, low-income farmers will continue to put pressure on the natural resources available to them. When these resources are used close to their "critical zones", the risk of irreversible damage to the resource base becomes an important consideration.

It is significant to note that most impoverished rural people live on inferior lands. Income policy therefore is important because social time preferences will tend to go down as income rises. It will then become more realistic to expect rural people to moderate their exploitation of soil resources.

Soil Conservation and the Energy Dilemma

Most of the energy needs of low-income rural and urban households in developing countries are satisfied through consumption of fuelwood. Income is negatively correlated with fuelwood consumption (Sfeir-Younis, 1983). Since energy for cooking, heating and lighting is a basic need, exploitation of existing forests has increased the degradation of many soils. When fuelwood is not available in adequate quantities, rural households will increase their use of such other energy sources as cowdung and agricultural residues. Since dung and residues can be used as fertilizer, their use as sources of energy will result in short-term declines in soil productivity.

Therefore, the satisfaction of energy needs will not only increase the soil erosion rate due to deforestation but will also decrease soil productivity.

Demographic Policy Dimensions

Population growth and demographic concentration are important factors in determining the rate at which soils are eroded. Over the years, Malthus' views on population and resources have often been revived. In simple terms, his view was that the limited quantity of resources would end up reducing the rate of population growth. Although Malthus' view has been contested on several grounds (for example, on grounds that population growth rates are an "independent variable"), it is important to recognize the areas which will be most affected by population pressure. Important losses in welfare may occur in areas where the use rates of soil resources exceed the "critical zone." Demographic policies may have to be put in place in these areas. Otherwise, resource management actions seem bound to fail.

Distribution of Wealth

The soil erosion process can have important distributional effects. The losses of soil suffered by one farmer may become gains of soil for another farmer, unnoticed losses of soil upstream may cause substantial losses in downstream nonfarm activities. Several distributional effects are worth mentioning here, all of which require policy actions. These are related to the distribution of welfare between:

1. The current generation, which bears the cost of soil conservation programs and the future generation, which benefits from such programs.
2. The economic agents (for example, farmers) within a project area and those outside that area.
3. Producers and consumers.
4. All of the farmers within the project area.
5. Landowners and farm operators.

These distributional effects are important in weighing policy decisions and establishing incentive systems. The answers to such questions as who is to benefit from conservation policies within this generation, or between generations, lie at the heart of controversial policy issues in welfare economics (for example, what is the appropriate discount rate?). The answers will also affect the way in which proposed projects are judged and selected.

Development and Conservation

To devise a policy framework that will achieve a proper balance between development and conservation is difficult and in the large majority of developing countries such a balance has seldom been achieved. It is often said that LDC policies are biased against conservation. This is not necessarily true. But if it is, there are important reasons for such a bias, (1) huge numbers of inhabitants living at subsistence levels and (2) frequent political instability, which encourages short-term planning.

The reality is that, even though soil conservation is a legitimate social goal, projects to control erosion must be looked at within the context of other development projects and a balance must be achieved. It is important for LDCs to consider all the potential uses of available natural resources.

Regional and Local Issues

Adoption Rates

One of the key elements in implementing soil conservation programs is farmers' adoption of practices that will conserve the soils. Adoption often can be assured only by intervention that ranges from the creation of public awareness of the problem to providing information, education, training and financial incentives. Public policy may simply limit the use of certain soil resources, or change the ways in which they are used.

Positive inducements include sharing the costs of land preparation and levelling, input or product subsidies and other monetary incentives. Negative inducements include higher taxes on certain activities, barriers to products and input markets and reduction of services to program nonparticipants.

In all cases, the adoption of adequate technologies and farming systems is the major requirement for successful implementation of soil conservation programs. The effects on farmers' income should be carefully analyzed in such situations.

Resettlement

In many instances, the deterioration of soils is so great that it becomes necessary to resettle inhabitants. The productivity of soils may be low and the risk of irreversible damages high. Resettlement of the people working those lands becomes a major social concern.

Thus, a major policy issue faced by several countries (for example, Indonesia) involves the potential resettlement of farmers in areas of potentially higher land productivity. Significantly large resettlement programs, if such an option is open to a country, have proven expensive and in some cases, very difficult to implement.

In countries facing land constraints, farmers often abandon the land (for example, Guatemala), migrating to the cities in search of employment. Although statistics are sparse, urban resettlement has also been reported to be expensive.

Investment Priorities

Investment decisions are guided by the expected benefits of such decisions. For a given set of benefits (for example, a given level of erosion rates), actual damage costs are an important element in deciding which policy to adopt. Given that the financial resources of governments are limited, governments will have to decide whether to allocate funds to control on-farm effects, or downstream effects, or both.

Research in the United States shows that the downstream damages caused by flooding are much greater than the value of land productivity foregone because of soil erosion. There has not yet been enough research to say whether the same is true of developing countries. But, if downstream effects do result in higher social costs, policy should initially be directed toward the control of sedimentation rates. Controlling sedimentation is less costly, in terms of organizational requirements, than controlling erosion on the farm and engineering solutions are much more readily available.

On the other hand, if off-farm effects are higher in value to the economy, investment policies and institutional changes would be of a different nature. Effectiveness in controlling the rate at which soils are eroding will also require important organizational improvements at the national, regional and farm levels.

Technology Policies

It is imperative that governments establish adequate policies for dealing with new technologies, particularly land-saving and land-conserving technologies. It is difficult to pinpoint an "optimal" technology policy, but certain things are clear. First, *the technology for controlling erosion is well-known.* Its adoption will be based mostly on a perception of its economic merits. But even in cases where the recommended technology is perceived to be potentially beneficial from an economic viewpoint, a series of activities need to be developed for a technical package to be adopted. For example, if contour ploughing is recommended, it is necessary for farmers to have access to ploughs or farm tools that will enable them to do the contours (for example, mobile plough base) particularly in areas characterized by steep slopes. Second, policies have often advocated the adoption of production or productivity-oriented technologies, without regard to the potentially negative impact on soils. Mechanization comes immediately to mind. Another example is the adoption of HYVs, which tend to exhaust the existing supply of soil micronutrients.

15

Compensation Issues

If the benefits of soil conservation projects were easily seen, we would expect farmers to undertake these measures themselves. The fact is that in most cases farmers do not see an indication that either a) the total benefits are not large enough to justify the expense, or b) some form of market failure exists. The thrust of this book is that market failure is common and therefore a wider analysis is needed to determine if a) is true or false. Intergenerational equity issues are closely related to this topic and are presented in Chapter 16.

Examined from another perspective, the difficulty in evaluating and implementing soil conservation projects arises from a *displacement of benefits and costs*. This displacement is both *spatial* (the upstream/downstream dichotomy referred to earlier) and *temporal* (the distribution of benefits and costs over time). This displacement in space and time leads to market failure. Intervention is therefore required to reach an economically and socially superior (acceptable) outcome.

A major problem in deciding the appropriate level of intervention is lack of data. For example, an analysis of soil erosion at numerous locations within a watershed can make it possible to estimate soil erosion costs more accurately. In this way, spatial distortion can be reduced. If the costs of soil conservation measures would be less than the erosion-caused damages, the soil conservation measures are economically justifiable. A subsidy or incentive scheme may be needed to encourage upstream farmers to undertake the conservation measures, and this compensation is economically justified under the Pareto improvement criteria.

Temporal distortions are much harder to analyze. If the full impact of soil erosion effects was felt in five or ten years, it could be assessed easily enough through discounting and accepted analytical procedures. But what about long-term effects: 20, 50 or 100 years? In this case, not only is there great uncertainty about what will happen at that distant time, but discounting also makes the values of costs or benefits very small when they are measured in today's terms. For example, $100 in benefits (or costs) 50 years in the future is today worth $60.80 at a one percent discount rate, $8.70 at a five percent discount rate and only $0.90 at a ten percent discount rate. Economic analysis, *per se*, does not handle these long-range effects well.

The use of a zero discount rate would give costs and benefits equal weight regardless of when they occur, but a zero discount rate is inefficient as a way of allocating current resources among different projects. Social welfare would probably decrease if all projects were evaluated at a zero discount rate. For these reasons, alternative ways of making intergenerational resource allocations are presented in Chapter 16.

In sum, economic analysis can give useful information about the appropriate levels of compensation for short-term soil conservation schemes. That is, spatial distortions can be overcome through economic analysis. But temporal distortions are more troubling and with respect to them there are limits to the role of economic analysis. For this reason, alternative criteria have been developed to help societies decide how much of today's soil resources should be preserved for future generations. Although individuals are mortal and therefore tend to make narrow and self-interested decisions, societies are longer-lived and should therefore have a broader set of concerns, including preservation of a larger amount of the soil resource base than might occur if decisions on preservation were left to individuals.

Compensation Schemes

The issue of compensation invariably arises in any discussion of soil conservation. There are two reasons for this. Firstly, new technologies and farm practices are key factors in explaining the success of many soil erosion control programs and these technologies and practices will often determine whether to continue to operate existing farms or resettle the farmers in other areas because of the risk of irreversible damage to their current land. Secondly, soil conservation programs, even when they do not require resettlement, will have an economic impact on farmers and other actors in the economy, with corresponding efficiency and equity implications. Fundamentally, soil conservation requires a change in the rules -- which theoretically transgresses the conventional efficiency grounds and requires the consideration of equity.

The long-term nature of the benefits of soil erosion control is what makes incentive or compensation schemes necessary. As stated earlier, the investments in soil conservation that farmers can afford may not be sufficient from a social standpoint. Therefore, compensation schemes may be required to give the present generation of farmers the incentive to adopt sometimes costly conservation practices. Many types of compensation schemes are being implemented in developing countries. These schemes are implemented in both formal ways through laws and regulations and in informal ways dictated by tradition. Very little has been written about the effectiveness of such programs in controlling erosion, or about their impacts on individual farmers and the economy.

Compensation schemes may include compensation to farmers for directly quantifiable losses in benefits or increases in costs, or for increases in risks and reductions in food consumption (if land is required to be put out of production). The instruments of compensation may operate through the marketplace or through direct government programs. The most important marketplace instruments include taxes, subsidies and wage and price support systems. The most common instruments of direct intervention involve compensation in kind (grain, in the case of the World Food Program), provision of services (for example, land levelling, construction of terraces) and direct income transfers (for example, food stamps). These incentives may be given to individuals or to collective units, such as villages. They may discriminate among regions, watersheds, cropping systems and the like.

Direct intervention programs provide farmers with concrete resources to balance out expected losses in yields or income because of the adoption of a soil erosion control program. Payments in kind may be equivalent in volume to the commodity that is displaced or foregone. Where resettlement of farmers is required, these schemes help in the transition period.

Compensation schemes raise both macroeconomic and microeconomic questions. The major macroeconomic questions concern the regional or national impacts of these compensation schemes. When farmers in a particular region are compensated, migration into the region may increase as outsiders try to obtain the benefits of the compensation too. But changes in the rate of migration, or in the allocation of labor and other inputs, especially under conditions of instability with regard to tenure or property rights, will increase land use pressures. When compensation programs are designed to benefit only one area, different types of soil conservation programs in other areas may be seriously affected. Farmers in the other areas may refuse to adopt conservation practices unless they too are compensated in some form.

Compensation schemes that subsidize the price of inputs (for example, fertilizer) or outputs may change market conditions. A common scheme is one to subsidize the price of capital inputs through lower interest rates, but important distortions may occur when these subsidies are given indiscriminately. If the price of fertilizer is subsidized, for example, farmers will be expected to use more fertilizer and fertilizer demand will rise. It does not necessarily follow that soil erosion will be reduced. On the contrary, larger applications of fertilizer may mask the effects of erosion.

The microeconomic issues involve such questions as how much? when? and what to compensate for? Compensation often means compensation for things where the market value is hard to establish. Determining the right compensation for farmers and their families compelled to resettle on new lands is a particularly difficult problem.

Many direct compensation programs involve various types of subsidies for labor. Certain reforestation schemes in foreign-financed projects in South Asia offer a good example of these. Reforestation of village-owned land (or public/state-owned land) is carried out by mobilizing family or wage labor. The labourers may get food, or a minimum wage, or part of the proceeds from forest harvests. The effectiveness of these programs has not been evaluated yet.

Other compensation programs include the provision of low-cost or free services inputs -- for example, seedlings. These items are free in the sense that farmers do not pay for them directly, although they may be paying for them indirectly through taxation. One of the main issues involved in supplying subsidies or incentives to the private sector is the potential waste. Free distribution of seedlings, for example, does not in itself ensure effective reforestation. Many of the seeds may be wasted if insufficient attention is paid to their survival and growth.

Farmers' Welfare, Transactions and Markets

Compensating farm households for losses in income or yields will inevitably have important welfare effects. Changes often occur through shifts in household consumption.

Incentive schemes may influence the relationships between individual farm households and the marketplace. For example, an incentive scheme that provides farmers with monetary compensation for losses in revenue or yield will have a favourable effect on farmers' budgets. If the incentive program involves the provision of food to farm households, the composition (mix) of consumption between food and non-food items may change. This, in turn, may increase the prices of both food and non-food products. On the supply side, compensation in kind may flood the market with compensated products to such an extent that the prices of some of these commodities may go down, thus reducing the net value of compensation.

Studies of the effects of compensation schemes on market prices and farm household income are badly needed.

Capital, Land and Labor Markets

Several compensation schemes augment, at low cost or no cost, the supply of physical or financial capital to farmers. Some programs, for example, involve the loan of tractors or earth-moving equipment to farmers. Others involve extension of credit to farmers to carry out soil erosion control programs. When incentives like these are administered indiscriminately, the net effects on soil conservation may be negative. The use of tractors, for example, may increase the area in deep tillage and have a negative effect on soil structure. The same applies to subsidized credit. Subsidized interest rates may induce overcapitalisation (for example, purchase of tractors), increasing the use of capital-intensive technologies.

Incentives also tend to affect land use by changing the income-generating possibilities of land. Land use distortions due to incentive schemes that show gains in the short term may not be sustainable in the long term if they increase erosion and if the subsidies must be perpetual.

Compensation schemes may have various effects on labour markets. Suppose, for example, that a soil erosion control program causes land productivity and profitability to increase. This increase in profitability, in turn, may result in greater use of labor-saving technologies that may intensify land use, thus worsening the erosion problem. A program in which the government pays for labor costs as an incentive for land conservation, on the other hand, may lead to the creation of elaborate labor-using projects, thus distorting the labor market.

Evaluation of WFP Interventions

Food aid is particularly useful for compensation purposes as, (1) attracting labor to labor-intensive conservation work, (2) giving labor an incentive to devote surplus time voluntarily to conservation activities, (3) shifting farmers' goals toward more rational use of land (given their constraints) and toward the adoption of new soil-conserving activities, and (4) compensating farmers for land temporarily left out of production to recover its productivity (particularly pasture lands). Food aid can also be instrumental in training and the proceeds from food sales to general consumers may be used to buy such non-food imports as fertilizer and machinery.

The World Food Program (WFP) has been assisting soil conservation projects since 1964. As of May 1984, WFP had assisted 43 soil conservation projects and 78 projects with soil conservation components. The total value of such assistance has been estimated at $243.3 million. The lessons learned from these projects can be summarized as follows:

1. Effectiveness in controlling erosion depends on land use planning and thus on directed and comprehensive approaches rather than half measures.
2. Priorities must be set, particularly with regard to the volume of aid (which determines both incentive levels and area coverage) and the duration of such programs.
3. Programs and techniques must be adapted to local conditions.
4. Food aid programs should be targeted toward socially and technically desirable projects.
5. Implementation of soil conservation projects must be accompanied by other types of interventions -- that is, the goal is not only to control erosion but also to improve farmers' welfare.
6. Present activities should, if possible, be designed in such a way that soil conservation programs can be maintained indefinitely into the future.
7. Government responsibilities must be established and government agencies expanded when necessary.

Compensation of Upstream Farmers by Downstream Users

Downstream water users (for example, irrigators, hydroelectric generating stations, water supply utilities) often realize that their output will be constrained if farming practices in the upper part of the watershed are not changed. To sustain downstream productivity, the owners or beneficiaries of such activities may decide to share the cost of soil conservation at the farm level.

The connection between upstream environmental degradation and adverse downstream effects is well-established, as are technical methods for reducing such damage. Frequently, the adoption of these methods imposes costs on land users upstream and provides benefits principally or exclusively to downstream users. Economic analysis can be used to determine whether downstream benefits justify upstream costs and to assist in designing schemes to compensate adversely affected groups.

Incentives and Their Economic Implications[1]

There is general agreement that insecurity of land tenure can lead to over-exploitation of land and a disinclination to invest in land improvements. There is no agreement however, on the merits to using subsidies, price intervention, direct regulation, or taxation to encourage soil conservation activities.

Subsidizing inputs. It has become customary to subsidize such farm inputs as fertilizer, irrigation water, land clearing activities and the growing of fodder crops. Fertilizer subsidies are supposed to have two positive effects, they increase farmers' incomes by allowing them to pay less per unit of fertilizer input and they increase land productivity by making it possible to add more nutrients to the soil. Very often however, fertilizer subsidies decrease the real opportunity costs of soil and nutrients, thus encouraging unsound cultivation practices. Since land productivity increases in the short term, farmers have no incentive to adopt land management practices to cope with soil degradation. Moreover, subsidizing fertilizer costs gives farmers an incentive to use marginal land or land whose productivity cannot be sustained over the long term.

Irrigation water subsidies tend to encourage farmers to use more water. Since farmers do not pay the true opportunity costs of the water (that is, water becomes a "cheap" input), farmers will irrigate areas that may not be suitable for intensive land cultivation. One of the results of this irrigation may be excessive levels of salinity in the soil.

Land clearing subsidies are often used to expand agricultural lands. These subsidies may be achieved through tax concessions or by providing land-clearing machinery and equipment at a nominal charge or no charge. But land clearing may have such negative consequences as destruction of vegetation, expansion of marginal lands very susceptible to erosion and increased salinity in secondary or dry lands.

Subsidization of fodder crops to mitigate drought effects is not uncommon. Cheaper fodder however, may cause the size of herds to remain constant or even increase. Increase in stock, in turn, may accelerate the process of land degradation through overgrazing.

Price intervention. Price supports (that is, prices higher than what the market would bear in the absence of intervention) will clearly attract economic resources to the production of price-supported crops. Although such interventions have a positive effect on farmers' incomes in the short term, several detrimental environmental effects may occur including expanded use of marginal lands, depletion of nutrients and micro-nutrients, direct erosion and avoidance of crop rotation systems.

Direct interventions. There are several ways in which governments can intervene to prevent further land degradation. These interventions may include the enactment of laws, rules, or regulations regarding different forms of land use, changing tenure and leasing systems and procedures, limiting the size of livestock or stocking rates, imposing specific agricultural patterns (for example, shifting land from cropping to grazing), or prohibiting the clearing of vegetation and the cutting of trees. The main obstacle to direct intervention is the problem of enforcement. It may simply be too expensive. Moreover, direct intervention requires sizable amounts of information that is often not available in developing countries. Moreover, direct intervention often lacks the flexibility needed as a result of the environmental specificity and tenure specificity of problems.

Subsidies for soil conservation practices. It is often argued that direct subsidies are necessary to give farmers enough incentive to adopt soil conservation practices. A typical form of subsidization occurs when governments absorb the cost of capital expenditures for conservation purposes (for example, terracing, land levelling). However, subsidies are not very effective in controlling erosion or sedimentation when, in addition to these expenditures, farmers have to change their land management practices. Perhaps direct subsidies are more effective in areas which are characterized by "simpler" farming systems. In this sense these subsidies will be more effective in many areas of developing countries. However, subsidies for capital expenditures may result in serious distortions of the investment allocation process. Subsidies of this type reduce farmer's opportunity costs, since farmers are not required to internalize the true costs of land degradation. In the long term, therefore, one may expect the rate of land degradation to increase.

Taxation. The previous policies are incentives (that is, "the carrot"). However, it is often proposed that taxes be used as disincentives (that is, "the stick"). The main objective of such taxes is to increase the cost of certain practices. As explained earlier however, erosion and sedimentation are difficult, if not impossible, to trace back to individual farms. Consequently, the costs of enforcement might be too high, compared to the damage that could be attributed to an individual farm. If farming systems are similar, a tax according to the size of farm may be possible.

A few conclusions can be drawn from the above discussion. First, even in cases where they may appear to be unconnected with land use and management, microeconomic policies may have important effects on conservation and depletion of soils. Second, assessment of macroeconomic and microeconomic interventions is needed to make sure that they do not defeat the very objectives they were designed to achieve. Third, such incentive or compensation schemes as food aid, subsidies, taxes and direct interventions may aggravate the land degradation problem, even though they improve farmers' incomes in the short term. Finally, the objective of incentive schemes should not only be to improve the attractiveness of soil conservation practices but also to internalize the negative effects of land degradation (that is, let those who produce the damages pay for the cost of those damages).

Notes

1. Based on Blyth and Kirby (1984).

16

Discounting and Intergenerational Equity[1]

Decision Criteria

A precise definition of the term "intergenerational equity" is very difficult and involves not only economic but ethical, moral and legal questions. A comprehensive treatment of these issues is beyond the scope of this book. We begin by stating a few questions that are often raised in discussions of intergenerational equity. These questions provide the basis for the discussion that follows.

(1) Should decisions on long-term investments in natural resources and management be made using discounting methods (and thus using the NPV or internal rate of return) as decision criteria?
(2) Are the discount rates commonly used by financial institutions in project analysis too high for this purpose?
(3) Should the rate of discount used in project analysis reflect the social time preference (that is, whether to use a natural resource today or tomorrow) of the generation presently making decisions, or of a future generation?
(4) Does the use of irreplaceable assets by this generation constitute "theft" from future generations?
(5) Is the present generation undervaluing future benefits (or costs) and therefore unnecessarily increasing the depletion rate of soil resources?
(6) Are current accounting procedures adequate for the future, or is this generation creating equity biases against future generations?

These questions reflect concern about the intertemporal use and management of soil resources and the answers are not easily found. Further, universal prescriptions that apply to all societies should not be expected. There is no consensus among scientists and policy makers on how these problems might be resolved. This section simply reviews the most important elements of the problems.

Discussions of intergenerational equity among economists usually focus on whether irreversible damages will occur from current exploitation of exhaustible resources. As stated earlier, soil is a composite resource including renewable as well as nonrenewable resources. Our concern is with the renewable components of soil resources that have a critical zone and with nonrenewable components.

The Economic Criterion

Investment decisions are often made on the basis of an economic criterion known as the maximization of Net Present Value (NPV). Under this criterion, decision makers should choose investments where the NPV is greater than zero and greater than any mutually exclusive investment, or where the Internal Rate of Return (IRR) is greater than the opportunity cost of capital. This criterion reflects some basic principles of welfare economics, the so-called potential compensation criteria, which are used to form social judgements by considering whether the gains of those who gain from a change would be large enough to compensate losers for their losses.

The traditional economic criterion for making decisions, particularly about investments that will affect future generations, is not the only valid one. As T. Page (1977b) points out, there is a "conservationalist" criterion which should also be taken into account.

The Net Present Value Rule

The use of the NPV rule, although widely accepted on grounds of efficiency, has often been criticized on the grounds of equity. Use of the NPV rule on grounds of efficiency is based on three assumptions: (1) competitive markets, (2) perfect information and (3) well-defined markets. When the NPV formulation is used to judge equity, the assumptions are that income distribution and preferences would remain the same with or without the project (Walton 1981).

The general implications of the NPV calculation are that the value of net incremental benefits (B minus C) becomes less significant in contributing to NPV in years further in the future. Benefits beyond, say, 15 years do not much affect the decision criterion at a discount rate of 10 percent -- the higher the rate the less the impact.

Some of the questions listed earlier refer to the extent to which one is able to assign proper values to environmental benefits and costs (including monetary costs) and reflect them in the numerator of the NPV function. Other questions pertain to the value of the discount rate: The higher the value, the less relevant are future benefits when this generation is making decisions with regard to resource use. Further, some people have suggested that the entire discounting framework is inadequate for making really long-term investment decisions. In response, many economists have suggested solutions that focus on the NPV function. For example, it has been suggested that much lower discount rates be used for environmental projects. Variations of this approach include multiple but declining discount rates (that is, the value of the discount rate is not independent of time). Others have suggested changing the rules for pricing benefits and costs to increase (or decrease) the value of the numerator. At this point however, many point out that the market only values "traditional' inputs and outputs (for example, fertilizer and rice) -- that is, that it does not value environmental goods and services. One solution, proposed by Krutilla (1967), was to take market values as well as option values (that is, what people are willing to pay to preserve an environmental service even if they do not actually use that service) into account. Several other combinations of the above mentioned approaches are found in the literature.

A common complaint is that economic analysis *undervalues* the environmental benefits that will occur in the future. The argument is that environmental goods and services will be scarce in the future (there will be fewer of them and more people with more money wanting them). Since traditional economic analysis assumes that all prices remain constant, these "appreciating" goods and services will be undervalued and therefore fewer of them will be provided than is socially desirable. The solution to this situation is quite simple, if the analyst feels that the *real* price (inflation-free) of an environmental good or service will increase over time, then the analysis should reflect this change in relative prices. As a result, future benefits (or costs) will be valued more highly. For example, some analyses of reforestation programs have assumed that tropical hardwoods would increase in price over time (perhaps 2.5 percent per year). This assumption is then factored into the analysis and increases the value of trees harvested 20 or 30 years in the future. This approach avoids the use of multiple or very low discount rates.

An example from Page (1977b) may be useful to illustrate the valuation and discounting problem. He states that these discounting methods can be viewed as dealing almost entirely with intergenerational efficiency and very little equity or fairness. He offers three intergenerational decision rules: (1)"straightforward discounting," (2) "almost anywhere dominance," and (3) "Pareto dominance."

Assume that present-day authorities are deciding between two land management schemes. One of them focuses on soil conservation now and has high start-up costs, but will supply goods and services at a lower cost when in place. The other focuses on continuous exploitation of soils and has lower start-up costs but will provide goods and services at a higher cost in the future.

How should society choose between these alternatives? If straightforward discounting is used, this generation will prefer the non-soil conservation (NSC) to the soil conservation (SC) project because the net benefits are concentrated near the present. If the "almost anywhere dominance" rule is used, the rule will decide for that project whose net benefits are higher almost anywhere than the other project's. In this case, the rule would favour the SC project. Use of the Pareto dominance rule will rank the NSC higher than the to SC for a given time frame and change the ranking for another time frame.

The Conservationist Criterion

Many conservationists argue that the discount rate for conservation projects should be much lower than the rate for other investment programs. Some have argued that society's discount rate is lower than private sector rates because of market imperfections, the role of government and microeconomic constraints. Others have argued that economists underestimate the value of environmental services. Finally, many have argued that one should superimpose an equity objective for aggregating welfare across generations and that the NPV criterion can be used to judge intertemporal efficiency but is not useful to assess intertemporal equity.

Page (1977b), based on Barnett and Morse, presents four rules advocated by proponents of the conservationist criterion:

1. "The regenerative capacity or potential of renewable resources (such as forests, grazing land, crop land, water) should not be physically damaged for destroyed".
2. "Renewable resources should be used in place of minerals, insofar as physically possible".
3. "Plentiful mineral resources should be used before less plentiful ones, insofar as physically possible".
4. "Nonrenewable resources should be recycled as much as possible."

In short, the conservationist criterion states that natural assets should be managed on the basis of sustainable yield. This term must be defined for each resource under consideration. In dealing with composite and interacting resources, sustainability is not a straightforward concept.

Page (1977b) outlines the "optimality" properties of the respective criterion. The optimality properties of the NPV criterion are:

1. *Administration.* The present value criterion is self-administering, since it recommends, by and large, what markets automatically do (Page, p. 191).
2. *Completeness.* The present value criterion is complete. By itself, it is in principle enough to yield a complete allocation of materials, goods and services in the economy (Page, p. 191).
3. *Intertemporal Fairness.* The present value criterion, or at least the selfish altruism version of it, favours the present generation over future generations (Page, p. 191).
4. *Efficiency.* The present value criterion is efficient in principle and efficiency is one of the chief virtues of the criterion. However, intertemporal efficiency is not always easy to analyze (Page p 191)
5. *Permanent Livability.* The present value criterion makes no guarantee about livability of the future (Page, p. 191).
6. *Intertemporal Consistency.* The present value criterion is inconsistent intertemporally (Page, p. 191-192).
7. *Conditional Evaluation.* The present value criterion is not suited to the evaluation of intertemporal externalities (Page, pp. 196)[2]

The optimality properties of the conservationist criterion are given as follows:

1. *Administration.* The conservation criterion is not self-administering.
2. *Completeness.* The conservation criterion is a partial criterion.
3. *Intertemporal Fairness.* The conservation criterion attempts to provide a fair use of the resource base intertemporally . . . is a partial criterion, it does not attempt to provide fairness in all its dimensions, but only in the use of the physical resource base.
4. *Permanent Livability.* The primary purpose of the conservation criterion is to keep the resource base intact.
5. *Intertemporal Consistency.* If each generation preserves the resource base for the next generation, then the resource base is

preserved for all generations . . . in that sense, the conservation criterion is intertemporally consistent.

6. *Conditional Evaluations.* The conservation criterion does not take into account explicitly the implicit trades which could benefit both present and future in the use of the resource base.
7. *The Conservation Criterion* . . . appears unnecessary when things are looking up, but it provides insurance against threats to the resource base decades hence. (Page, pp. 198 and 199)

Reconciliation of the two sets of criteria requires the use of basic theory of "contracts" with regard to intertemporal choices. The definition of "reasonable" contracts is at the heart of the current controversy.

Haveman (1977) proposes that reasonable contracts would require comparing the present value of net benefits with the present value of future damages to the environment. This could be done on the basis of the following assumptions: (1) environmental damages (present and future) are known, (2) the value of these damages is also known, and (3) the discount rate reflects preferences between present and future consumption. If the NPV is greater than the present value of damages, there is a potential increase in the welfare of the community. Freeman (1977), based on Haveman's argument, states that discounting methods are useful in dealing with intergenerational equity issues because they enable planners to estimate the amount of compensation to be paid to future generations.

Hartwick (1978b) has argued (and proved mathematically) that efficient paths (or contracts) are characterized by constant per capita consumption when population is stationary. The maintenance of constant per capita consumption forces policy makers to look at future costs -- that is, at what costs will the future generation be able to maintain such per capita consumption? Crosson (1983) expands this argument within the context of soil conservation. He states that "intergenerational equity in management of agricultural resources can be defined as avoidance of secular increases in real costs of producing food and fibre. The necessary condition for achieving this is that total agricultural productivity rises in step with demand. Some loss of productivity of the land is consistent within this if there are compensating increases in the productivity of non-land resources. At the margin, advances in technology and management may substitute for the loss in productivity of the land, thus avoiding increase in production costs (p. 25)." The most important point raised by Crosson is that dealing with intergenerational equity does not imply reducing losses in land productivity to zero. Planners would need to focus on all major factors affecting production costs.

Ferejohn and Page (1978) have proposed the use of the "dominance rule" to solve intergenerational issues. On the assumption that there is an infinite number of generations and therefore an infinite time horizon: Four conditions for the social choice of contracts are proposed. These conditions need to be satisfied in order to reflect the preferences of individuals in society. "They are: (1) transitivity (T), (2) unanimity (U), (3) independence of irrelevant alternatives (IIA), and (4) non-dictatorship (N). However, the NPV formulation follows another principle (the stationary principle), which states that if in the first time period programs A and B are identical but A is preferred to B because of subsequent benefits, then the social ordering will still rank A as preferred to B after the first period is completed." (Walton, 1981, p. 243).

The conditions that would be established either by the dominance rule or the stationary principle cannot be satisfied simultaneously. Any combination of these conditions will result in social choices weighted on the side of this generation (or else a subgroup of generations will dominate). Because of these problems, Walton suggested what he calls a "partitioned" approach. This consists of dividing the time horizon of the project into different parts. He suggests time segments of 5 years, on the grounds that the worth or a project will become apparent within that time horizon. "If a project benefits the current generation but all the risks and costs are borne by future generations, this pattern of gains is made apparent by the partitioning approach." (Walton, p. 246)

A comparison of the partitioning approach with the NPV approach suggests two points: (1) while "who pays" and "who benefits" are indistinguishable in the NPV formulation, the partitioning approach can provide such information, and (2) while the NPV reduces compensation to "one number," the partitioning approach allows one to focus on which generation should be compensated.

The partitioning approach is interesting, but one could argue that the NPV approach is also part of the partitioning approach. Moreover, partitioning the time horizon of a project is arbitrary. Why every 5 years and not every 2 or every 10 years? Under certain conditions this might not be an important issue. For certain projects however, the partition period could be a crucial issue.

Where Do We Go from Here?

The presentation in this chapter may create a feeling of uneasiness. The topic itself is difficult, the presentation is brief and evaluation of any of the criteria is difficult. Comparisons of some of the non-economic criteria are practically impossible to make, since no attempts to apply them have been made. Also, the assumptions underlying each criterion reveal such differences in perspective that they are practically noncomparable.

An even more fundamental issue is to determine whether selecting these criteria should be a basic objective of conservation policy or simply a constraint. In particular, should the selection of a criterion on intergenerational equity be the objective of resource management policy, relegating economic efficiency (or a given level of economic efficiency) to the level of a constraint? Although this is a critical question, there is no straightforward answer. The emotional aspects of intergenerational equity should not be allowed to cloud our judgement. There are costs as well as benefits associated with soil conservation. Some societies may decide to accept higher levels of soil erosion than others because they need to satisfy other demands (for example, public education, health care, housing). There are no easy or absolute answers, economic analysis can provide decision makers with valuable information about options and tradeoffs.

The intention here was to show that decisions about the exploitation of the natural environment cannot be based exclusively on economic criteria and also that some criticisms of the NPV criterion are unwarranted.

The need for investment in erosion control programs is no longer questioned. The questions, rather are how much investment should take place and how should the costs be shared. With regard to cost-sharing, the literature shows that it is well accepted today that some form of compensation should be given to farmers who implement soil conservation practices.

Just as societies have trouble seeing from one generation to another, individual farmers may have difficulties planning from year to year, especially if they do not own the land they are farming. Pressures for immediate consumption and uncertainty about receiving the benefits of soil conservation, reinforce a common tendency to "mine" the soil, seek short-term gains and ignore longer-run consequences. This uncertainty frequently frustrates attempts to implement soil conservation measures. Farmers often see the costs of foregoing present consumption as too high, even if incentives are given. Government implementation of soil conservation measures (for example, terracing, land levelling) may be necessary in these cases.

Controversy surrounds the question of how much erosion this generation should allow. Clearly, economic criteria offer a wealth of courses of action to decision makers in developing countries. Crosson (1983) and Sfeir-Younis (1983) point out that economic and social conditions in many parts of the world rule out any efforts to reduce soil erosion to zero.

Concern for future generations will take the form of erosion control programs that do not unduly increase the soil costs to future generations, whether in terms of food and fibre production (Crosson, 1983, pp. 26-27) or in terms of other environmental services.

This will require comprehensive development programs. As explained earlier, erosion control programs should not be seen only as agronomic programs. Development programs should include sound policies. Income and property rights policies will be extremely important.

In sum, we suggest the following option. Since intergenerational equity issues are extremely difficult to quantify, it may be prudent to err on the conservative side -- that is, to invest more in soil conservation than can be justified by short-run economic criteria. An economic analysis normally considers the "costs" of present erosion rates whether they occur upstream or downstream. If these costs are prevented or reduced, society benefits. The magnitude of these benefits can then be compared to the costs of soil conservation to determine how much to invest. An alterative approach would be to inflate the value of these benefits by some factor (10 percent? 25 percent? 40 percent?) and use the higher figures to determine project expenditures. Although society would invest more in soil conservation than would normally be economically justified from an individual farmer's viewpoint, this larger investment would help preserve more soil resources for future generations.

The problem with this "conservative" approach is that it would divert resources from other uses to soil conservation. This may prove difficult unless there is an explicit government policy to invest in soil conservation. This approach has the advantage of providing a base for establishing the total level of resource use (in this case, of benefits identifiable in a wider economic analysis). A base avoids the economically inefficient and perhaps unattainable, use of goals like "zero erosion" or "no more than five tonnes/ha/year." As should be clear by now, the question of how much investment in soil conservation is justifiable is not answerable by either economic or physical factors alone.

Notes

1. Drawn from Page (1977).
2. See details from Page (1977).

17

Strategy and Research Needs

A Framework for Decision Making

The objectives of this book are to create greater awareness of the soil conservation problem, to provide new frameworks for the economic analysis of soil conservation projects and to draw attention to the most important issues. However, many subjects addressed by this book will demand important policy decisions.

The main objectives of this chapter are, to identify the most important policy issues, to establish a framework for both policy decisions and strategies and, to list topics for further research. Since the emphasis is on the decision making process, this chapter suggests a framework that will enable planners to put different policy prescriptions into a more cohesive and comprehensive perspective.

Three levels of decision-making are distinguished here, the policy level, the institutional level and the operating level.[1]

- The *policy level* focuses on the formulation, interpretation and execution of sectoral goals and objectives.
- The *institutional level* focuses on the market and nonmarket incentive arrangements for decision making.
- The *operating level* focuses on decision and actions concerning the allocation and use of natural resources.

These levels of decision-making are interrelated and therefore the optimal rules and regulations, as well as the decisions on allocation and use, will depend on how society decides on its development objectives. Several elements of a conservation strategy are outlined in relation to these three levels of decision-making.

Policy Level

At the national level, the following elements of a soil conservation strategy should be taken into account. These relate mainly to the role of national entities.

1. Establishment of a resources management policy, including land use, as part of an overall development plan,
2. Achievement of a national consensus on a soil conservation strategy,
3. Integration of soil conservation objectives with conditions and constraints in other sectors of the economy,
4. Creating awareness of the problem motivating people to participate in soil conservation programs,
5. Development and adoption of a technology policy for agricultural development,
6. Definition of investment selection criteria in light of the distribution of benefits and costs of soil conservation programs,
7. Achieving short-term stabilization to accomplish long-term objectives,
8. Definition of the roles to be played by the public and private sectors,
9. Allocation of funds and skilled personnel to initiate a national agricultural research effort to develop comprehensive soil conservation programs,
10 Establishment of the proper balance between public and private sector investments,
11 Establishment of the principle that future generations will not be forced to pay the social costs of current but inadequate development plans,
12 Formulation of reasonable targets and goals.

At the international level, the following elements of a soil conservation strategy should be taken into account. These relate mainly to the role of International Financial Entities (IFEs):

1. Coordination among IFEs, at both the country and regional levels, to avoid conflicts at the policy and program levels during implementation.

2. IFE mobilization of resources to expand programs geared to ameliorating and solving environmental problems.
3. IFE policy dialogue should include environmental concerns in a comprehensive way, where these concerns are an integral part of country planning and policy formulation.

Institutional Level

The main elements of soil conservation strategy at this level of decision-making are:

1. Establishment of market and nonmarket institutions to control soil degradation and create the necessary incentives to carry out soil conservation programs.
2. Formulation of incentive and compensation schemes.
3. Assessment of the potential impacts of these incentives on the allocation and use of soil resources at the national, regional and farm levels.
4. Coordination of government interventions to prevent fragmented action.
5. Assessment of macro and microeconomic policy instruments in terms of both traditional aggregates (for example, employment income, foreign exchange) and environmental related aggregates (for example, land use, water quality).
6. Establishment of soil conservation standards acceptable not only to farmers but also to other sectors affected by land degradation.
7. Enactment of appropriate legislation and regulations, with appropriate enforcement mechanisms.
8. Improvement of the existing and organizational structure at the national (for example, countries) and local levels (for example, farmer organizations).

Operating Level

At the operating level, the following elements of a soil conservation strategy should be taken into account:

a. Improvements in the design of soil conservation programs. Experience shows that effective strategies require changing the basic "unit of account" for investment decisions (for example, from the individual farm to the watershed). Both upstream and downstream effects should be considered in designing the unit of account.

b. Assessment of the "critical zone" in soil at the country and local levels.
c. Creation of multi-disciplinary teams and unified organizational structures to improve the implementation of soil conservation programs.
d. Changing the product mix in agriculture to be compatible with new land husbandry policies.
e. Creation of monitoring and evaluation units.
f. Expansion of extension, training and education programs.
g. Development of any ancillary activities necessary to avoid constraints on technology and resource allocation.
h. Improvements in the marketing and distribution of agricultural outputs.

Future Research Needs

This book is only the first step in outlining the nature and magnitude of soil erosion problems. Much more research is needed. This research should cover a large array of topics and it should be funded at levels that allow the collection and processing of adequate data. In addition, research should focus on both the macro and the micro economic aspects of soil erosion. Most research today is microeconomic and does not provide a framework for policy decisions. Economic, organizational, institutional and policy aspects are rarely considered.

Research in the following areas is needed:

1. *Soil Erosion and Productivity.* Very few studies go beyond estimating soil losses and it is therefore difficult to assess the economic merits of conservation programs.
2. *Elements Determining Project or Program Success.* Many soil conservation techniques are well-known and success in using them is dependent upon organizational and institutional factors.
3. *Assessment of Downstream Effects.* It seems clear from the available data that the value of downstream damages caused by erosion is quite substantial. Research should focus on assessing the sedimentation status of major reservoirs in the world, the kind and extent of damage caused by floods and changes in water supply and water quality.

4. *Regional Integration of Soil Conservation Programs.* In many instances, erosion is a consequence of livestock grazing or energy needs (for example, over-exploitation of existing forests to satisfy the demand for fuel-wood). Research into the integration of soil conservation programs at the regional level is needed. A potentially effective method of integration would be comprehensive research on farming systems.
5. *Organizational Arrangements.* Studies should be conducted to assess the most viable organizational schemes at the national as well as the local level. Organizational fragmentation, a major characteristic of soil conservation programs, is decreasing the effectiveness of many potentially successful conservation programs.
6. *Impact on Nonmarket Incentives.* The impact of alternative regulatory, tenure and property right systems on soil conservation is almost always unknown.
7. *Creation of a Data Bank.* Countries need to establish procedures to make maximum use of the experimental data and sectoral studies that have already been carried out.
8. *Assessment of Compensation Schemes.* Farmers are being compensated in several ways to carry out soil conservation practices and more research is needed to assess the financial and economic impacts of these schemes.
9. *Contribution of Soils to Development.* Agricultural development models and macroeconomic policy focus more on aggregates (for example, income, employment, foreign exchange) than on the potential implications of use and misuse of natural resources. Most current research is production-oriented and does not take into account the role of soil resources in the development of agriculture in the economy.

Concrete Actions

The following is a list of concrete actions that can be carried out by national and international institutions.

National Programs

1. Assign more importance to macroeconomic policy on soil conservation.
2. Increase financial allocations for conservation work.
3. Change the unit of account in planning and decisionmaking.

4. Establish compensation schemes.
5. Strengthen soil conservation organizations.
6. Establish low-cost and replicable programs.
7. Create a viable institutional framework.
8. Implement program in fragile areas.
9. Increase funding on soil/water conservation research.

International Financial Institutions

a. Change investment selection criteria.
b. Strengthen environmental work in relation to sector and country strategies.
c. Establish new training programs.
d. Coordinate aid programs.
e. Integrate actions with those of WFP, WMD, FAO, UNESCO and UNDP.
f. Change focus of some international research centers in favour of environmental concerns.

Epilogue

How Did We Get Where We Are Now?

The nature and magnitude of the erosion problem in many countries is such that action is urgently needed. Erosion is a natural phenomenon, but imbalances have been created by human actions. Worsening erosion and sedimentation will impair the achievement of long-term development objectives.

Erosion and sedimentation are caused by many different kinds of human actions and a comprehensive evaluation of these actions across almost every sector is needed. It is not very useful to concentrate exclusively on actions taking place at the farm level when, for example, road construction is not done properly and causes erosion.

We got where we are now partly because of lack of integration between sectoral policies and macroeconomic policies. Macroeconomic policies on such things as prices, export incentives, foreign exchange earnings and the like, that might appear to have no connection with land degradation are indirectly having a substantial impact on soil quantity and quality.

It is essential that the economic aspects of the land degradation problem are well understood. This will enable policy makers and development institutions to avoid major mistakes. However, land degradation is not only an economic problem. It is a social problem as well and policy makers should understand that the effects extend beyond this generation and that it causes are not simply the actions of individual farmers.

Land degradation is also a social, demographic, physical and perhaps most important of all, a political problem. Difficult decisions must be made to avoid irreversible damages to the basic foundations of development, people and national resources. Preventative actions are certainly more appropriate than curative actions. Thus, strategies to deal with land degradation require firm political commitments now.

Investment decisions, a major subject in this book, are only one part of the dilemma. Clearly, a development strategy must determine the proper balance between private and public investment. But decision makers must change their present perspective and integrate decisions on natural resource conservation with food production, research and extension and education.

Where Do We Go from Here?

Even at the risk of confronting tough decisions and making mistakes, the worst possible decision would be to do nothing. Decision makers should accept the fact that they may fail several times before seeing the light of success.

After reviewing the vast and voluminous literature on the subject, one gets the impression that several determinants of success can be clearly identified:

1. In many instances, very minor changes (accompanied by a favourable policy environment) can check the rate at which land is being degraded. In other words, *soil conservation practices are not necessarily expensive*.
2. *Farmers must participate in the decision making process*. Clearly, a lack of farmer participation is responsible for the poor sustainability of conservation decisions and practices.
3. Management *objectives and policies must be developed* and these must be accompanied by practical *guidelines and standards*.
4. Income and food-related policies must accompany conservation policies. In other words, conservation policies *cannot be enacted or put into place in isolation*.
5. Effectiveness in policy or program implementation depends upon *knowledge and information*. Lack of these two ingredients will

result in decisions which, in many instances, lack even common sense.

6. Investments and policies should be judged not only on their economic merits but on other grounds, sometimes including *ethical or moral* grounds.
7. *Replicability and area coverage* are two basic determinants of projects success. But the implications of this are several. Government and financial institutions must understand that in order to control land degradation, it is imperative to deal with vast areas of a country, even at the risk of supporting economically "unprofitable" projects.
8. Investments and policies should be oriented toward *internalizing the negative effects* of human actions, whether of farmers or other actors in the economy. Those responsible for the damages should somehow be made aware of the costs of correcting the damages -- in some cases, by helping to pay the costs.

How Do We Get Where We Want to Go?

To achieve success in a reasonable time, it is imperative to take an integrated approach across sectors, across decision-making levels and across private and public sector boundaries.

A move toward rational exploitation of the natural resources available to agriculture and other sectors must be guided by a clear set of management objectives. Without these objectives, investment and policy changes will be carried out in a vacuum. Several objectives can be singled out, (1) the need to look into soil and water quality as well as quantity changes, (2) the recognition of demographic problems, (3) assessment of conservation/development tradeoffs, and (4) definition of an action program to lead the way into the future.

With regard to soil quantity and quality, a basic set of management objectives must be followed to avoid irreversible damages, to reverse degradation problems in the medium term, to reduce erosion to tolerable levels and to encourage the retention of prime agricultural land. With regard to water quantity and quality, investments and policies should be designed to minimize the adverse effects of organic wastes, to reduce pollution from excessive nutrients and salinity, to minimize levels of toxic pollutants and to reduce sedimentation effects.

It is important to identify areas where the population places stress on the natural resource base. In many instances it may be necessary to protect fragile land while using it to produce food or fibre, while living with certain negative impacts -- at least in the short term.

Coping with the challenges ahead will not be an easy task. Countries as well as financial institutions will need to establish a comprehensive conservation agenda, set reasonable targets and focus on achievements.

Let us remember that poor land makes people poor.

Notes

1. See Yamauchi and Onoe (1983), pp. 133-139 and Ciriacy-Wantrup, (1968).

Bibliography

Abiodun, A.A. 1973. "Water Resources Projects in Nigeria and the Hydrological Data Employed in Their Planning and Development," in UNESCO/WMO/IAHS, *Design of Water Resources Project with Inadequate Data*. Proceedings of the Madrid Symposium, June, pp. 21-33.

Academia Nacional de Agronomia y Veterinaria, Instituto Nacional de Technologia Agropecuaria. 1979. "La Erosion del Suelo en la Cuerca del Plata," IDIA No. 379-384 July-December pp. 1-96.

Agency for International Development. 1983. Foreign Disaster Assistance. By Evaluation Technologies Inc Arlington VA for USAID Office of US Foreign Disaster Assistance *Various Country Profiles*. May.

Agricultural Research Service, USDA. 1972. *Present and Prospective Technology for Predicting Sediment Yields and Sources Proceedings of the Sediment Yield Workshop*. USDA Sedimentation Laboratory, Oxford, Mississippi, November 28-30, ARS-S-40 Washington, D.C. USDA June.

Agricultural Research Service, USDA. 1976. Proceedings of the Third Federal Sedimentation Conference, March 22-25. Denver, Colorado.

Ahmad, Y.J. ed. 1982a. *Environmental Guidelines for Irrigation in Arid and Semi-arid Areas*. UNEP Environmental Management Guidelines No. 2, Nairobi; UNEP.

Ahmad, Y.J. ed. 1982b. *Environmental Guidelines for Watershed Development*. UNEP Environmental Management Guidelines No. 3, Nairobi, UNEP.

Ahmad, Y.J. ed. 1982c. *Analyzing the Options Environmental Cost-Benefit Analysis in Differing Economic Systems*. UNDP Studies, 5, Nairobi, UNEP.

Ahmad, Y.J., P. Dasgupta and K.G. Maler, eds. 1984. *Environmental Decision Making, Vols. 1 and 2*. London, Hodder and Staughton.

Alauddin, M and C. Tisdell. 1989. "Biochemical Technology and Bangladeshi Land Productivity, Diwan and Kallianpur's Analysis Reapplied and Critically Examined," *Applied Economics UK*. Vol, 21 6 June, pp. 741-760.

American Society of Agricultural Engineers. 1977. *Soil Erosion and Sedimentation Proceeding of the National Symposium on Soil Erosion and Sedimentation by Water*. December 12-13, 1977 Palmer House, Chicago, Illinois. St Joseph, Michigan, American Society of Agricultural Engineers.

American Soil Association/Soil Science Society of America. 1982. *Determinants of Soil Loss Tolerance*. Madison, Wisconsin, ASA/SSSA.

Anderson, F., A. Kneese, O. Reed, S. Taylor and R. Stevenson. 1977. *Environmental Improvement Through Economic Incentives*. Baltimore, The Johns Hopkins University Press.

Anderson, J.C., E.O. Heady, and W.D. Shrader. 1963. "Profit Maximizing Plans for Soil Conserving Farming in the Spring Valley Creek Watershed in Southwest Iowa," Research Bulletin 519. Agricultural and Home Economics Experiment Station, Iowa State University, Ames, Iowa, July.

Anderson, J.E. and H.C. Bunch. 1989. "Agricultural Property Tax Relief, Tax Credits, Tax Rates and Land Values," *Land Economics*. Vol. 65, 1 Feb, pp. 13-22.

Andrus, C. and D. Southgate. 1986. "The Economics of Erosion Control in a subtropical Watershed, Comment/Reply," *Land Economics*. Vol. 62, 3 Aug, pp. 329-332.

Arid Lands Information Center. 1981. *Draft Environmental Report on Various Countries* U.S. Man and the Biosphere Secretariat, Department of State, Washington, D.C.

Arnoldus, H.M.J. 1977a. "Methodology used to Determine the Maximum Potential Average Soil Loss due to Sheet and Rill Erosion in Morocco," in *Assessing Soil Degradation*. FAO, pp. 99-124.

Arnoldus, H.M.J. 1977b. "Predicting Soil Losses Due to Sheet and Roll Erosion," in *Guidelines for Watershed Management*. FAO, pp. 39-48.

Ateshian, K.H. 1976. "Comparative Costs of Erosion and Sedimentation Control Measures in *Proceedings of the Third Federal Inter-Agency Sedimentation Conference*. March 22-25, Denver, Colorado, pp. 2-13-2-23.

Avery, D.R. 1979. "Soil Erosion, Conservation and Non-commercial Energy is the Third World, Area Annotated Bibliography," Ann Arbor, School of Natural Resources, University at Michigan, April.

Babu, R, R.C. Barsal, and M.M. Srivastava. 1978. "Effect of Bunding on Runoff and Peak Discharge in Agriculture Watershed in Doon Valley," *Indian Journal of Soil Conservation*. Vol. 6, No. 2, October, pp. 89-93.

Balachandran, C.S, Fisher, P.F. and M.A. Stanley. 1989. "An Expert System Approach to Rural Development, A Prototype TIHSO," *Journal of Developing Areas*. Vol. 23 No. 2 Jan, pp. 259-270.

Balba, A.M. 1983. "The Aswan High Dam and its Impact on Egyptian Agriculture," *Outlook on Agriculture*. Vol. 12, No. 4, pp. 185-190.

Balci, A.N.U. 1979. "Rapport national du comite MAB de Turquie," *Hommes, Terre et Eaux*. Vol. 9, No. 3 January-February, pp. 125-142.

Barbier, E.B. 1990. "The Farm-Level Economics of Soil Conservation, The Uplands of Java," *Land Economics*. Vol. 66, No. 2, May, pp. 199-211.

Barnett, A.P. and others. 1972. "Soil and Nutrient Losses in Runoff with Selected Cropping Treatments on Tropical Soils," *Agronomy Journal*. Vol. 64 May-June pp. 391-395.

Barry, P.J. and Robison, L.J. 1986. "Economic Versus Accounting Rates of Return for Farm Land," *Land Economics*. Vol. 62 4 Nov pp. 388-401.

Bartelli, L.J. 1979. *Mexico Rainfed Agricultural Development Project Soil and Water Conservation*. Consultant Report, Washington, World Bank, September.

Batie, S.S. and R.G. Healy eds. 1980. *The Future of American Agriculture as a Strategic Resource*. Washington, D.C. The Conservation Foundation.

Batie, S.S. 1983a. *Soil Erosion, Crisis in America Croplands*. Washington, D.C. The Conservation Foundation.

Batie, S.S. 1983b. *Priority Issues in Soil Conservation*. Paper prepared for Philip Morris Workshop, Agriculture in the 21st Century, April 1-13, Richmond, Virginia.

Bauder, J.W. *et al*. 1981. "Continuous Pillage, What It Does to the Soil," *Crops and Soils Magazine*. December, pp. 15-17.

Baumann, R.V, E.O. Heady, and A.R. Aandahl. 1955 "Costs and Returns for Soil Conserving Systems of Farming on Ida-Monona Soils in Iowa", Research Bulletin 429. Agricultural Experiment Station, Iowa State College, Ames, Iowa, June.

Baumgartner, A. 1970. "Water and Energy Balances of Different Vegetation Covers," in World Water Balance Proceedings of the Preceding Symposium July Vol. 3. See IASH/UNESCO/WMO.

Baumol, W.J. 1950. "The Community Indifference Map, A Construction", *Review of Economic Studies*. Vol. 15, No. 3 pp. 189-197.

Baumol, W.J. 1965. "On the Social Rate of Discount," *American Economic Review*. Vol. 59, No. 5 December, pp. 930.

Baumol, W.J. 1972. "On Taxation and the Control of Externalities," *American Economic Review*. Vol. 62, pp. 307-22.

Baumol, W.J and W.E Oates. 1975. *The Theory of Environmental Policy Externalities, Public Outlays and the Quality of Life*. Englewood Cliffs, New Jersey, Prentice-Hall, Inc.

Beattie, W.D. 1979. *Evaluation Socio-Economic de Actividades de Conservacions y Rehabilitacion de Tierras en el Valle de Tarija-Bolivia*. FAO Project PCT/BOL/8802, Tarija, Bolivia, FAO May.

Bekkali, A and others eds. 1979. *Hommes, Terre et Eaux*. Special Issue "L'Erosion et L'Ameragement des Bassins Versont Dans les Pays Mediterraneans," Vol. 9, No. 30 January-February, pp. 175.

Bell, C. and P. Hazell. 1980. "Measuring the Indirect Effects of an Agricultural Investment Project on Its Surrounding Region," *American Journal of Agricultural Economics*. Vol. 62, pp. 75-86.

Bell, R.D. 1981. "Terrace Cultivation, A Challenge to Modern Technology," *Span*. Vol. 24, No. 3, pp. 131-133.

Benbrook, C. (nd) Review of the Yield-Soil Loss Simulator. Memorandum of the Council on Environmental Quality, Executive Office of the President, Washington, D.C.

Bender, L.D. 1987. "The Role of Services in Rural Development Policies," *Land Economics*. Vol. 63 1 Feb pp. 62-71.

Bennett, H.H. 1939. *Soil Conservation*. New York, McGraw Hill.

Bennett, H.H. 1960. "Soil Erosion in Spain," *Geographical Review*. Vol. 50, No. 1 Jan, pp. 59-72.

Bennett, C. 1975. *Man and Earth's Ecosystem*. New York: John Wiley.

Ber Salam, B. 1980. "Arid-zone Forestry - where there are no forests and everything depends on trees," *Unasylum*. 32 128, pp. 16-18.

Berg, R.D. and D.L. Carter. 1980. "Furrow Erosion and Sediment Losses on Irrigated Cropland," *Journal of Soil and Water Conservation*. Vol. 35, No. 6 November-December pp. 267-270.

Berggren, R. 1975. *Economic Benefits of Climatological Services*. WMO Report No. 424, Geneva.

Berglung, S.H. and E.L. Michalson. 1981. "Soil Erosion Control in Idaho's Cow Creek Watershed, An Economic Analysis," *Journal of Soil and Water Conservation*. Vol. 36, No. 6, May-June pp. 161.

Bhalla, S.S. 1988. "Does Land Quality Matter? - Theory and Measurement," *Jrnl of Development Economics Netherlands*. Vol. 29 1 Jul pp. 45-62.

Bhalla, S.S. and Roy, P. 1988. "Mis-Specification in Farm Productivity Analysis, The Role of Land Quality," *Oxford Economic Papers UK*. Vol. 40 1 Mar pp. 55-73.

Bhatt, P.N. 1977. "Losses of Plant Nutrients through Erosion Process - A Review," *Soil Conservation Digest*. Vol. 5, No. 1 April, pp. 37-46.

Bidard, C and Woods, J.E. 1989. "Taxes, Lands and Non-Basics in Joint Production," *Oxford Economic Papers UK*. Vol. 41 4 Oct pp. 802-812.

Binswanger, H.P. 1975. "The Use of Quality Between Productivity Profit and Cost Functions in Applied Econometric Research, A Didactic note," *Occasional Paper No. 10*. Economics Department ICRISAT Hydrabad India July.

Binswanger, H.P. and M.R. Soserzweig. 1982. "Behaviour and Material Determinants of Production Relations in Agriculture," *Research Unit, Agriculture and Rural Development Department Report No. 5*. World Bank Washington D.C. June.

Birch, A. *et al*. 1983. "Toward Measurement of the Off-site Benefits of Soil Conservation," Economic Research Science, *Agriculture Economics Report*. No. 341, May.

Biswas, A.K. ed. 1977. *Water Management and Environmental in Latin America*. Oxford, Pergamon Press.

Blyth, M.J. and M.G. Kirby. 1984. "The Impact of Government Policy on Land Degradation in the Rural Sectors," Paper presented at the 54th ANZAAS Congress, Australia National University, Canberra, Australia, May 14-18.

Bogess, W, J. Mchrann, M. Beohlje, and E.O. Heady. 1979. "Farmland Impacts of Alternative Soil Loss Control Policies," *Journal of Soil and Water Conservation*. Vol. 34, No. 4 July-August, pp. 177-183.

Bognetteau-Verlinder, E. 1980. *Study on Impact of Windbreaks in Majjia Valley, Niger*. Bouza, Niger, CARE/Wageningen Agricultural University, February.

Bohm, P. 1973. *Social Efficiency, A Concise Introduction to Welfare Economics*. New York, John Wiley & Sons.

Bonsu, M. 1980. "Assessment of Erosion Under Different Cultural Practices on a Savanna Soil in the Northern Region of Ghana," in *Soil Conservation Problems and Prospects*. R.P.C. Morgan, ed. 1980. pp. 247-253.

Bonvallot, J.A.H. 1977. "Causes et modalites de l'erosion dans le lbassin versont inferieur de l'Oued El-Hadjel Tunisie Centrale", in *Erosion and Sediment Transport in Inland Waters*. IASH, pp. 260-268.

Botero, L.S. 1982a. *Watershed Rehabilitation, Torrent Control and Flood Protection*. "Lecture Notes prepared for FAO/Finland Training Course on Watershed Management for Africa, August 9-28, Nairobi, Kenya.

Botero, L.S. 1982b. *How a Watershed functions, Concepts and Characteristics*. Lecture Notes prepared for FAO/Finland Course on Watershed Management for Africa, August 9-28, Nairobi, Kenya.

Bouchafra, A. 1979. "Le cas du LOUKKOS et le sous-bassin de NEFZI," *Hommes, Terre et Eaux*. Vol. 9 No. 30 January-February pp. 55-64.

Bourcart, J. 1957 *L'Erosion des Continents*. Paris, Libraire Armand Colin.

Bouvard, M. 1983. "Ouvrages de derivation et transports solides," *La Houille Blance*. No. 3/4, pp. 247-254.

Bouwes, N.W. and R.R. Schneider. 1978. "Estimating Water Quality Benefits," *Economic Research Service*. Working Paper, Number 55, December.

Bouzaher, A., Braden, J.B. and Johnson, G.V. 1990. "A Dynamic Programming Approach to a Class of Nonpoint Source Pollution Control Problems," *Management Science*. Vol. 36 1 Jan pp. 1-15.

Bower, D.K. 1972. "Sediment Control," *Agricultural Engineering*. Vol. 53, No. 7 July, pp. 17-19.

Braden, J.B. 1980. *Research on Soil Erosion and Water Quality*. Department of Agricultural Economics, December. Urbana-Champaign, University of Illinois.

Braden, J.B., Lawrence, B.A., Tampke, D and Wu, Pei-Ing. 1987. "A Displacement Model of Regulatory Compliance and Costs," *Land Economics*. Vol. 63 4 Nov pp. 323-336.

Breimyer, H.F. 1991. "New Farm Law Saves More Dollars but Less Soil," *Challenge*. Vol. 34 3 May/Jun pp. 52-53.

Brobaker, S. 1977. "Land-The for Horizon," *American Journal of Agricultural Economics*. Vol. 59, No. 5 December Proceeding Issue pp. 1037-1043.

Brokensha, D. and B. Riley. 1978. *Forest, Foraging, Fencing and Fuel in a Marginal Area of Kenya*. Paper prepared for the U.S. Agency for International Development, African Bureau Firewood Workshop. June 12-14, Washington D.C.

Bromley, D.W. 1980. *The Political Economy of Private Resource. Use Decisions Affecting Soil Productivity*. Paper prepared for workshop on "Soil Transformation and Productivity," sponsored by the U.S. National Academy of Sciences - National Research Council, October 16-17.

Bromley, D.W. 1982. "Rights of Society versus Landowners and Operators," H. G. Halcrow and et al. eds, *Soil Conservation Policies, Institutions and Incentives*. Soil Conservation Society of America, Iowa, USA, pp. 219-232.

Brooke, C. 1967. "Food Shortages in Tanzania," *The Geographical Review*. Vol. 57, No. 3 July, pp. 333-57.

Brooks, K.N., H.M. Gregorsen, E.R. Bergung, and M. Tayaa. 1982. "Economic Evaluation of Watershed Projects - An Overview Methodology and Application," *Water Resources Bulletin*. Vol. 18, No. 2 April pp. 245-250.

Brookshire, D, A. Randall and J. Stoll. 1980. "Valuing Increments and Decrements in Natural Resource Service Flows," *American Journal of Agricultural Economics*. Vol. 62, pp. 478-88.

Brown, L. and E.C. Wolf. 1984. *Soil Erosion, Quiet Crisis in the World Economy*. Worldwatch Paper 60. Washington, Worldwatch Institute.

Brown, L.H. 1973. *Conservation for Survival, Ethiopia's Choice*. Addis Ababa, Haile Sellassie I University.

Brown, L.H. 1978. *The Worldwide Loss of Croplands*. Worldwatch Institute, Paper 24.

Brown, L.H. 1979. "Where Has All the Soil Gone?," *Mazingira*. No. 10 pp. 61-68.

Brown, L.R. 1981. "World Population Growth, Soil Erosion and Flood Security," *Science*. Vol. 214 September 27, pp. 995-1002.

Brown, L.R. 1989. "The Grain Drain, The Waning of Food Security," *Futurist*. Vol. 23 4 Jul/Aug pp. 9-16.

Brown, M. 1979. *Farm Budgets, From Farm Income Analysis to Agricultural Project Analysis*. Baltimore, The Johns Hopkins University Press, for World Bank.

Bunyard, P. 1980. "Terraced Agriculture in the Middle East," *The Ecologist*. October, November, pp. 309-316.

Buras, N. "The Cost-Effectiveness of Water Resources System Considering Inadequate Hydrological Data," UNESCO/WMO/LAHS, pp. 649-700.

Burnett, E., B.A. Stewart, and A.L. Black. 1983. "Regional Effects of Soil Erosion on Crop Productivity - Great Plains," paper prepared for Soil Erosion and Crop Productivity Symposium, March 1-3, Denver, Colorado.

Burns, M. 1973. "A Note on the Concept and Measure of Consumers' Surplus," *American Economic Review*. Vol. 63 pp. 335-44.

Burt, O. 1981. "Farm Level Economics of Soil Conservation in the Palouse Area of the Northwest," *American Journal of Agricultural Economics*. Vol. 63, pp. 83-92.

Burz, J. 1977. "Suspended-Load Discharge in the Semi-arid Region of Northern Peru," in UNESCO and IAHS, *Erosion and Solid Matter Transport in Inland Waters*. pp. 269-277.

Butlin, J.A. ed. 1981. *Economics of Environmental and Natural Resources Policy*. Westview Press, Boulder, Colorado.

Carpenter, R.A. 1983a. "Ecology in Court and Other Disappointments of Environmental Science and Environmental Law," *Natural Resources Lawyer*. Vol. 15, No. 3 pp. 573-595.

Carpenter, R.A. 1983b. *Natural Systems for Development, What Planners Need To Know*. New York, MacMillan Publishing Company.

Carter, D.L., M.J. Brown, and J.A. Bondurant. 1976. "Sediment - Phosphorus Relations in Surface Runoff from Irrigated Lands," in *Proceedings of the*

Third Federal Inter-Agency Sedimentation Conference. March 22-25, Denver, Colorado, pp. 3-41 - 3-52.

Carvajalino J.L.J. 1948. "Los Ingenieros Agronomos Rehabilitaran la Agricultural Colombiana," *Agric. Trop. Bogota*. Vol. 4 pp. 8-12.

Casas, R.R. and M.J.R. Zaflanella. 1979. "El Manchoneo de los Suelos en la Region Este de la Provincia de Santiago del Estero", *IDIA*. No. 379-384 July-December pp. 103-122.

Castellanos, V. and J.L. Thomas. 1980. *Application of Multiple - Use Research on Watershed in Honduras*. Paper presented at the IUFRO/MAB Conference, Research on Multiple-Use of Forest Resources, May 20-23, Flagstaff, Arizona.

Castle, E., Kelso, M., Stevens, J. and H. Stoevener. 1981. "Natural Resource Economics. 1946-1975," *A Survey of Agricultural Economics Literature*. Vol. 3, L. Matin gen. ed. Minneapolis, University of Minnesota, pp. 395-500.

Castro, J.P. 1984. "Watershed Management of the Lower Agno River Project," Paper presented at the East-West Center Workshop on The Management of River and Reservoir Sedimentation in Asian countries. May 14-19. Environment and Policy Institute, East-West Center, Honolulu, Hawaii.

Centre for Science and Environment. 1982. *The State of India's Environment 1982*. A Citizens Report. New Delhi, Centre for Science and Environment.

Charbouni, Z. and others. 1979. "Rapport National Tunisien," *Hommes, Terre et Eauz*. Vol. 9, No. 30 January-February, pp. 91-124.

Chardrasekharan, C. 1981. *Causes and Consequences of Land Degradation*. Paper prepared for an informal Workshop on the Economic Aspects of Land Conservation under Humid and Semi-Humid Tropical Conditions, June 30-July 3, Rome.

Chipman, J. and J. Moore. 1980. "Compensating Variation, Consumers' Surplus and Welfare," *American Economic Review*. pp. 933-49.

Choil, H. and C. Coughenoue. 1979. *Socioeconomic Aspects of No-Tillage Agriculture. A Case Study of Farmers in Christian County, Kentucky*. University of Kentucky, June.

Chong, K. 1974. "Checking Erosion on the Loess Land," *China Reconstructs*. Vol. 23, No. 4, April, pp. 22-27.

Choudbury, G.R. 1984. "Management of Sediment in Bangladesh," Paper presented at the East- West Center Workshop on The Management of River and Reservoir Sedimentation in Asian Countries, May 14-19. Environment and Policy Institute, East-West Center, Honolulu, Hawaii.

Christensen, D.A. Turhollow, E.O. Heady and B. English. 1983. "Soil Loss Associated with Alcohol Production From Corn Grain and Corn Residue," *CARD Report 115*. Ames, Center for Agricultural and Rural Development, Iowa State University.

Christensen, L. ed. 1982. *Perspectives on the Vulnerability of U.S. Agriculture to Soil Erosion*. An Organized Symposium, American Agricultural Economics Association Summer Meeting, Logan, Utah.

Christiansson, C. 1981. *Soil Erosion and Sedimentation in Semi-Arid Tanzania.* Scandinavian Institute of African Studies, Uppsala and Department of Physical Geography, University of Stockholm.
Christy, L.C. 1971. *Legislative Principles of Soil Conservation.* FAO Soils Bulletin 15 Roma, FAO.
Chunkao, K. "Sediment Transport from Main Rivers to the Gulf of Thailand," abstract in *Proper Summarized International Conference on Soil Erosion and Conservation.* p. 79.
Ciancio, O. 1979. "effect du type de boisement sur l'ampleur de l'erosion dars les ensembles hydrologiques de Calabre," *Hommes, Terre et Eaux.* Vol. 9, No. 30 January-February, pp. 159-161.
Ciriacy-Wantrup, S.V. 1938a. "Soil Conservation in European Farm Management," *Journal of Farm Economics.* Proceedings Number February pp. 86-101.
Ciriacy-Wantrup, S.V. 1938b. "Economic Aspects of land Conservation," *Journal of Farm Economics.* Vol. 20, No. 2 May pp. 462-473.
Ciriacy-Wantrup, S.V. 1941. "Economics of Joint Costs in Agriculture," *Journal of Farm Economics.* Vol. 23, No. 4 November pp. 771-818.
Ciriacy-Wantrup, S.V. 1942. "Private Enterprise and Conservation," *Journal of Farm Economics.* Vol. 24, No. 1 February, pp. 75-96.
Ciriacy-Wantrup, S.V. 1945. "Discussion," *American Economic Review.* Vol. 35, No. 2 May pp. 130-133.
Ciriacy-Wantrup, S.V. 1946a. "Review of Food or Famine, The Challenge of Erosion by Ward Shepard," *Political Science Quarterly.* Vol. 61, No. 2 June pp. 259-262.
Ciriacy-Wantrup, S.V. 1946b. "Resource Conservation and Economic Stability," *Quarterly Journal of Economics.* Vol. 60, pp. 412-452.
Ciriacy-Wantrup, S.V. 1946c. "Administrative Coordination of Conservation Policy," *Journal of Land and Public Utility Economics.* Vol. 22, No. 1 February pp. 49-58.
Ciriacy-Wantrup, S.V. 1947. "Capital Returns from Soil Conservation Practices," *Journal of Farm Economics.* Vol. 29, No. 4, part 2 November pp. 1181-1196.
Ciriacy-Wantrup, S.V. 1951. "Water quality, A Problem for the Economist," *Journal of Farm Economics.* Vol. 43, No. 5 December pp. 1133-1144.
Ciriacy-Wantrup, S.V. 1959. "Philosophy and Objectives of Watershed Development," *Land Economics.* Vol. 35, No. 3 August pp. 211-221.
Ciriacy-Wantrup, S.V. 1964. "The 'New' Competition for Land and Some Implications for Public Policy," *Natural Resources Journal.* Vol. 4, No. 2 October pp. 252-267.
Ciriacy-Wantrup, S.V. 1967. "Water Policy and Economic Optimizing, Some Conceptual Problems in Water Research," *American Economic Review.* Vol. 57, No. 2 May pp. 179-189.
Ciriacy-Wantrup, S.V. 1968. *Resource conservation economics and policies.* Uni of Calif, Div. of Ag. Sci. Ag. Exp. Station, 395pp. 3rd Edition.

Ciriacy-Wantrup, S.V. 1969. "National Resources in Economic Growth, The Role of Institutions and Policies," *American Journal of Agricultural Economics*. Vol. 51, December pp. 1314-1324.

Ciriacy-Wantrup, S.V. 1971. "The Economics of Environmental Policy," *Land Economics*. Vol. 47, No. 1 February pp. 36-45.

Ciriacy-Wantrup, S.V. and R.C. Bishop. 1975. "Common property as a concept in natural resources policy," *Natural Resources Journal*. 15 Oct , 713-728.

Clark, C. 1982. *Flood*. Time-Life Books, Alexandria, Virginia.

Clarke, H.R and A.K. Dragun. 1989. *Natural resource accounting; East Gippsland case study*. Australian Environmental Council.

Clarke, H.R. 1989. "Combinatorial Aspects of Cropping Pattern Selection in Agriculture," *European Jrnl of Operational Research Netherlands*. Vol. 40 1 May 5, pp. 70-77.

Claus, R.J, Large, D.W. and Claus, K.E. 1987. Irrigated Agriculture, Economic, Environmental, Disposal and Legal Problems *Appraisal Jrnl*. Vol. 55 3 Jul pp. 406-418.

Clawson, M, ed. 1964. *Natural Resources and International Development*. Baltimore, Johns Hopkins Press.

Clawson, M. 1974 "Economic Trade-Offs in Multiple-Use Management of Forest Lands," *American Journal of Agricultural Economics*. Vol. 56, No. 5, December, pp. 919-926.

Cliado, B. 1979. "Commataires sur les methodes utilisees pour letude l'erosion et l'Aminagement des bassins Versants," *Hommes, Terre et Eaux*. Vol. 9, No. 30 January-February, pp. 163-167.

Colombani, J. 1977. "Effets sur les transports solides des ouvrages hydrauliques en Afrique du Noid," in *Erosion and Sediment Transport in Inland Waters*. IAHS, pp. 295-300.

Comite Francais des Grands Barrages. 1976. "Problemes de Sedimentation Dans Les Retenues," Paper presented at the 12th Congress of Large Dams, Mexico, pp. 1177-1208.

Comite Francais des Grands Barrages. 1980. "Control de L'Alluvionnement de Retenues Quelques Example Types," Paper presented at the Seminaire International D'Experts Sur Le Devasement Des Retinues. Tunisia, July 1-4, pp. 537-562.

Committee on Selected Biological Problems in the Humid Tropics. 1982. *Ecological Aspects of Development in the Humid Tropics*. Washington D.C. National Academy Press.

Conclin, L.R. and others. 1978. "Economics of On-Farm Methods of Controlling Sediment Loss from Surface - Irrigated Fields," *Bulletin No. 584*. Moscow, Idaho, University of Idaho, Agricultural Experiment Station, June.

Cook, C.C. and Grut, M. 1990. "Agroforestry in Sub-Saharan Africa", *Finance & Development*. Vol. 27 3 Sep pp. 46.

Cooke, H.J. 1983. "The Struggle Against Environmental Degradation - Botswana's Experience," *Desertification Control*. No. 8, June pp. 9-15.

Cooper, C. 1981. *Economics Evaluation and the Environment*. London, Hodder and Stoughton.

Corbel, J. 1959. "Vitesse de L'erosion," *Zeitschrift for Geomorphologie*. Annals of Geomorphology pp. 1-28.

Corway, F.J. 1979. *A Study of the Fuelwood Situation in Haiti*. Report prepared for USAID Mission to Haiti.

Cory, D.R. Gum, and W. Martin. 1981. "Use of Paasche and Laspesyres Variations to Estimate Consumers' Welfare Change," *Agricultural Economic Research*. April.

Cotner, M.L. 1969. "A policy for Public Investments in Natural Resources," *American Journal of Agricultural Economics*. Vol. 51, No. 1 February pp. 87-99.

Coughlin, R.E. and others. 1980. *National Agricultural Lands Study Executive Summary The Protection of Farmland*. Washington, U.S. Government Printing Office, December.

Couvreur, M. 1977. "Une Nouvelle Forme de Gestion des Terres Recuperees au Maroc, La Societe de Development Agricole SODEA", *Mediterranee*. Vol. 29, pp. 97-102.

Cowen, R. 1983. "Transoceanic Dust Blows Out of Agro-Asian Deserts," *Christian Science Monitor*. April, 21.

Creager, W.P. 1929. *et al. Engineering for Dams*. Vol. 1. General Design, John Wiley and Sons, Inc. New York.

Crosson, P. 1975. "Environmental Consideration in Expanding Agricultural Production," *Journal of Soil and Water Conservation*. Vol. 30, No. 1, January-February, pp. 23-28.

Crosson, P. ed. 1982. *The Cropland Crisis*. Baltimore, Johns Hopkins University Press.

Crosson, P. 1983. *Soil Erosion in Developing Countries, Amounts, Consequences and Politics*. Resources for the Future. Seminar paper. Department of Agricultural Economics, University of Wisconsin, November 15, Madison, Wisconsin.

Crosson, P. and S. Brubaker. 1982. *Resource and Environmental Effects of U.S. Agriculture*. Washington, D.C. Resources for the Future.

Crosson, P. and J. Miranowski. 1982. "Soil Protection, Why, by Whom and for Whom?" *Journal of Soil and Water Conservation*. Vol. 37, No. 1 January-February pp. 27-29.

Currie, J.J. Murphy and A. Schmitz. 1971. "The Concept of Economic Surplus and its Use in Economic Analysis," *Economic Journal*. Vol. 81, pp. 741-99.

Curry-Lindahl, K. 1974a. "Conservation Problems and Progress in Equatorial African Countries," *Environmental Conservation*. Vol. 1, No. 2 Summer, pp. 119-121.

Curry-Lindahl, K. 1974b. "Conservation Problems and Progress in Northern and Southern Africa," *Environmental Conservation*. Vol. 1, No. 4 Winter, pp. 263-270.

Curtis, R.M. Jr. 1976. "Erosion and Sediment Yield in New Mexico," in *Proceedings of the Third Federal Inter-Agency Sedimentation Conference*. March 22-25, Denver, Colorado, pp. 181-190.

Daly, H.E., H.S. Burness, R.G. Cummings and R. Norgaard, R. 1986. "Thermodynamic and Economic Concepts as Related to Resource-Use Policies, Comment/Reply/Synthesis," *Land Economics*. Vol. 62 3 Aug pp. 319-328.

Daly, H.E. 1991. "Towards an Environmental Macroeconomics," *Land Economics*. Vol. 67 2 May pp. 255-259.

Dandy, F.E. and W.A. Champion. 1975. *Sediment Deposition in U.S. Reservoirs Summary of Data Reported through*. U.S. Department of Agriculture Miscellaneous Pub. No. 1362, Washington, D.C. U.S.D.A.

Das, D.C. 1977. "Soil Conservation and Practices and Erosion Control in India - A Case Study", in *Soil Conservation and Management in Developing Countries*. FAO.

Das, D.C., B.K. Mukherjee, and R.A. Karl. 1980. *Quantification of Multiple Benefits through mini Case Studies in the River Valley Projects Catchments*. Paper presented to the First National Symposium on Soil Conservation and Watershed Management, March, Dehra Dun, India.

Dasgupta, P. and G. Heal. 1974. "The Optimal Depletion of Exhaustible Resources," *Rev of Economic Studies*. Vol. 41, Supplement, pp. 3-28.

Dasgupta, P. and G. Heal. 1979. *Economic Theory and Exhaustible Resources*. Cambridge, Cambridge University Press.

Dasgupta, P. and K.G. Maler. 1981. "Selected Case Studies in the Applications of Cost Benefit Analysis to Environmental Resource," UNEP, IG. 29/71, September.

Daugherty, H.E. 1973. *Environmental Considerations in El Salvador Recommendations for national Action*. A translation of "Conservacion Ambiental en El Salvador. Recomendaciones para un Programa de Accion Nacional," N. Y, York University.

David, E.L. 1968. "Lakeshore Property Values, A Guide to Public Investment in Recreation," *Water Resources Research*. Vol. 4 pp. 697-707.

David, L. and M. Fogel. 1982. *Interactive Analysis of River Basin-Water Project Development*. Paper presented to the Fourth World Congress on Water Resources, September 5-9, Buenos Aires, Argentina.

Davies, D.P., D.J. Eagles, and J.B. Finney. 1972. *Soil Management*. Suffolk, Farming Press.

Day, H.J. 1973. *Benefit and Cost Analysis of Hydrological Forecast*. WMO Report No. 341, Geneva.

Day, R.H. 1978. "Adoptive Economics and Natural Resources Policy," *American Journal of Agricultural Economics*. May, pp. 276-283.

de Camino V.R. 1979. *Estimacion de Costos y Beneficios de la Reforestacion y la Conservacion de los Suelos en el Noroeste de Honduras*. Working Documents No. 2 Project No. PNVD - FAO-HOI/77/066. Tegucigalpa,

Honduras, FAO/UNDP/Corporacion Hondurena de Desarrollo Forestal, Octubre.

de Caraaff, J. 1981. *Watershed Development Activities Proposed for the Kingston Watersheds; Analysis of Costs and Benefits*. FAO Project Working Paper WP 19/Econ. JAM/78/006 GCP/JAM/-005/NOR Kingston, Government of Jamaica/UNDP/FAO, December.

Dean, G.W., E.O. Heady, S.M.A. Husain, and E.R. Duncan. 1958. "Economic Optima in Soil Conservation Farming and Fertilizer Vie for Farms in the Ida Monora Soil Area of Western Iowa," *Research Bulletin 455*. Agricultural and Home Economics Experiment Station, Iowa State College, Ames, Iowa January.

Deaton, A. and J. Buellbauer. 1981. *Economics and Consumer Behaviour*. Cambridge, Cambridge University Press.

Dekker, R. 1961. "Sediment Transport Measurements and Computations in the Niger," Commission for Technical Cooperation in Africa South of the Sahel CCTA, *Inter-African Conference in Hydrology*. Nairobi, pp. 256-265.

Delwaulle, J.C. 1973. "Increasing unproductiveness of Africa south of the Sahara," (Drought, erosion, migration) Fre. *Bois Forets Tropical*. Vol. 149, pp3-20 May/June.

Demmak, A. 1980. "L'Experience Algerienne en Matiere De Lutte Contre L'Envasement Des Barrages," Paper presented at The Seminaire Inernational D'Experts sur Le Devasement Des Retenues. Tunisia, July 1-4.

Departement des Sciences du sol de l'Institute Agronomique Hassan II. 1972. "Optimisation du Dessalage par Eau Saumatre des Sois du Tafilalt," *Hommes, Terre et Eaux*. No. 3-2, pp. 67-86.

Devarajan, S, and A.C. Fisher. 1981. "Hotelling's 'Economics of Exhaustible Resources', Fifty Years Later," *Journal of Economic Literature*. Vol. 19 March, pp. 65-73.

Diamond, P. and D. McFadden. 1974. "Some Uses of the Expenditures Functions in Public Finance," *Journal of Public Economics*. Vol. 3 pp. 3-21.

Dixon, J. 1983. *Policy Options for fuelwood Development*. Paper presented at the Benefit-Cost Analysis Workshop co-sponsored by the Office of Environment and the East-West Environmental and Policy Institute, June 20-25. Seoul, Korea.

Dixon, J. and K.W. Easter. 1985. "Economic Aspects of Integrated Watershed Management", paper presented at Workshop on Integrated Watershed Management. Honolulu, East-West Center. January 7-12.

Dixon, J. and M.M. Hufschmidt, eds. 1984. *Economic Valuation Techniques for the Environment, A Case Study Workbook*. Honolulu, East-West Center. Process.

Dixon, R.G. 1982. *Proyecto Para la Conservacion de Recursos Naturales Suelos Hidrologia y Bosques en la Cuenca de Rio Paute*. Ministerio de Agricultural y Ganaderia, Quito, Ecuador.

Dixon, R.K. 1988. Forest Biotechnology Opportunities in Developing Countries *Jrnl of Developing Areas*. Vol. 22 2 Jan pp. 207-218.

Doll, J.P, R. Widdows and P.D. Velde. 1983. "Research Review - The Value of Agricultural Land in the United States, A Report on Research," *Agricultural Economics Research*. Vol. 35, No. 2 April pp. 39-44.

Dongelmans, L. 1980. *Analysis Financiero de Reforestacion para Lena y de Eucaliptus en Terrazas*. Working Document No. 6, Project No. PNUD-FAO-HON/E77/006. Tegucigalpa, Honduras, FAP/UNDP/Corporacion Hondurena de Desarrollo Forestal, September.

Donovan, G. 1969. *Evaluation of Public Investments in Agricultural Projects in New Zealand, Practical Notes*. Department of Agriculture, Hamilton, New Zealand. July.

Dooms, P.L. 1972. "La Maitrise des Crues au Maroc. Contribution de l'ANAFID a la Revue Mondiale de l'ICID," *Hommes, Terre et Eaux*. No. 3 pp. 5-50.

Doran, J.W. 1982. "Tilling Changes Soil," *Crops and Soils Magazine*. Vol. 34, No. 9 August- September pp. 10-12.

Doublet, J.M. and F. Falloux. 1982. *Land-Related Issues in Agricultural and Rural Development, A Review of the Bank Approach*. Unpublished World Bank Report, Washington, D.C. October.

Douglas, I. 1967. "Man, Vegetation and the Sediment Yields of Rivers," *Nature*. Vol. 215 August 26, pp. 925-928.

Douglas, I. 1968. "Erosion in the Sungei Gombak Catchment, Selangor, Malaysia," *Journal of Tropical Geography*. Vol. 26 June pp. 1-16.

Dragun, A.K. 1980. "Natural Resource Management, Common Property and the Prisoners Dilemma," CRES Working Paper R/WP 50.

Dragun, A.K. 1981. "Coastal Management, Market Failure and the Role of Government," in *Offshore Structures*. The National Committee on Coastal and Ocean Engineering of the Institution of Engineers, Australia Perth, November.

Dragun, A.K. 1983a. "Common property and economic policy," Canberra, CRES, ANU. CRES Working Paper R/RP/22.

Dragun, A.K. 1983b. "Social decision making in the natural environment," Canberra, CRES, ANU. CRES Working Paper R/WP/25.

Dragun, A.K. 1983c. "Surface coal mine rehabilitation, Social costs and policy options," Canberra, CRES, ANU. CRES Working Paper R/WP/12.

Dragun, A.K. 1983d. "The practical application of valuation methodologies," In *Economics and Environmental Policy, The Role of Cost Benefit Analysis*. Dept of Home Affairs and the Environment, AGPS.

Dragun, A.K. 1983e. "Externalities, property rights and power," *Journal of Economic Issues*. September.

Dragun, A.K. 1984a. "The environmental impacts of surface mine development in the Hunter region of New South Wales," *International Journal of Environmental Studies*. Vol. 23, 179-189.

Dragun, A.K. 1984b. "Water management problems in a rapidly developing area, The Hunter region of New South Wales. *International Journal of Water Resources*. Vol 2 (4).

Dragun, A.K. 1985. "From remedial taxes to direct property rights approaches to externality," *Journal of Economic Issues*. Vol XIX (1) March.

Dragun, A.K. 1986. *Research review on the microeconomics of water management*. Australian Water Research Advisory Council, Canberra, AGPS.

Dragun, A.K. 1987. "Property rights in economic theory", *Journal of Economic Issues*. Vol XXI (2) June, 859-868.

Dragun, A.K. 1989. "The Economics of Institutional Change", in proceedings of "Property rights and institutional design," Conference at Lincoln College, Christchurch New Zealand, Feb 10th, Ministry of the Environment, New Zealand.

Dragun, A.K. and V. Gleeson. 1989. "From water law to transferability in New South Wales," *Natural Resources Journal*. Vol 29(3), 645-662 Summer.

Dudal, R. 1982. "Land Degradation in a World Perspective," *Journal of Soil and Water Conservation*. Vol. 37, No. 5 September-October, pp. 245-249.

Dudal, R. 1984. "Inventory of Major Soils of the World, With Special Reference to Mineral Stress Hazards," FAO, Rome. (Unpublished).

Dumsday, R. and W. Seitz. 1982. *A System for Improving the Efficiency of Soil Conservation Incentive Programs*. Urbana-Champaign, University of Illinois, Department of Agricultural Economics.

Dunne, T. 1976. *Studying Patterns of Soil Erosion in Kenya*. No. 7, Department of Geological Sciences and Quaternary Research Center, University of Washington, Seattle.

Dunne, T. 1979. "Sediment Yields and Land Uses in Tropical Catchments," *Journal of Hydrology*. Vol. 42, No. 3/4 pp. 281-300.

Durum, W.H. *et al*. 1970. "World-Wide Runoff of Dissolved Solids," International Abstracts of Scientific Hydrology, *General Assembly of Helsinki*. Publication No. 51, Belgium, pp. 618-628.

Dworkin, D.M. ed. 1974. *Environment and Development*. Indianapolis, SCOPE.

Dyhr-Nielsen, M. 1982. *Long-Range Water Supply Forecasting*. WMO Report No. 587. Geneva.

Dyke, P.J.W. 1982. "Erosion Productivity Impact Calculator EPIC, A Demonstration of Process Modelling for Policy Analysis," United States Department of Agriculture Economic Research Service Natural Resource Economics. *Division Working Paper No. 3*. April.

Earvin, C.A. and D.E. Ervin. 1982. "Factors Affecting the Vie of Soil Conservation Practices," *Land Economics*. Vol. 58, No. 3 August pp. 277-292.

East-West Environment and Policy Institute. 1981. *Benefit-Cost Analysis of Natural Systems and Environmental Quality Aspects of Development*. Chapter 6, East-West Environment Policy Institute Publication.

Eckholm, E. 1975. "The Deterioration of Mountain Environment," *Science*. Vol. 189, p. 764 September 5.

Eckholm, E. 1976. *Losing Ground, Environmental Stress and World Food Prospects*. New York, W. W. Norton & Company, Inc.

Edwards, J. 1980. *Some Interesting Equivalencies in the Analysis of Consumer Surplus in a three Commodity World.* Unpublished note, Oregon State University, April.

Egypt, Government of. 1977. *Aquatic Weed Problems in Egypt.* Abstract of paper presented at United Nations Water Conference, March, Mar Plata Argentina. E/CONF/. 70/ABSTRACT 20, September 13.

Eicho, C. and W. Lawrence, eds. 1964. *Agricultural in Economic Development.* New York, McGraw-Hill.

El-Swaify, S.A. and E.W. Dangler. 1983. "Erodibilities of Selected Tropical Soils in Relation to Structural and Hydrologic Parameters", in *Soil Erosion, Prediction and Control The Proceedings of a National Conference on Soil Erosion.* Soil Conservation Society of America, pp. 105-114.

El-Swaify, S.A., E.W. Donylor, and C.L. Armstrong. 1982. "Soil Erosion by Water in the Tropics," *Research Extension Series 024.* College of Tropical Agriculture and Human Resources University of Hawaii, Hitar. December.

Elfring, C. 1983. "Land Productivity and Agricultural Technology," *Journal of Soil and Water Conservation.* Vol. 38, No. 1 January-February, pp. 7-9.

Elkins, D.M. 1981. "Conservation Tillage, Is it the Key," *Crops and Soils Magazine.* December, pp. 15-17.

Elwell, H.A. 1978. *Soil Loss Estimation System for Southern Africa.* Department of Conservation and Extension, Zimbabwe.

Elwell, H.A. and M.A. Stocking. 1982. "Developing a Simple Yet Practical Method of Soil- Loss Estimation," *Tropical Agriculture. (Guildford)* Vol. 59, No. 1 January pp. 43-48.

Elwell, H.A, and M.A. Stocking. 1984. "Estimating Soil Life-Span for Conservation Planning," *Tropical Agriculture.* Vol 61, No. 2 April pp. 148-150.

Enabor, E.E. 1980. *Economics of Tropical Forest Resources Conservation.* Paper presented at the International Forestry Seminar, November 11-15, Kuala Lumpur, Malaysia.

Ervin, D.E. 1982. "Soil Erosion Control on Owner-Operated and Rented Cropland," *Journal of Soil and Water Generation.* Vol. 37, No. 5, September-October, pp. 285-287.

Ervin, D.E. and Dicks, M.R. 1988. "Cropland Diversion for Conservation and Environmental Improvement, An Economic Welfare Analysis," *Land Economics.* Vol. 64 3 Aug pp. 256-268.

Ervin, D.E. and R.A. Washborn. 1981. "Probability of Soil Conservation Practices, Missouri," *Journal of Soil and Water Conservation.* Vol. 36, No. 2 March-April pp. 107-111.

Falloux, F. 1983. *Sierra Leone, Analysis of Land Issues and Recommendations.* Unpublished World Bank Report, Washington, D.C. January.

Fan, Shou-Shoun. 1976 "The Role of Sediment Problems in Hydroelectric Development," *Proceeding of the Third Federal Inter-Agency Sedimentation Conference.* March 22-25, Denver, Colorado, pp. 4-149 - 4-161.

FAO/Bureau of Reclamation, United States Department of the Interior. 1974. *Soil Survey in Irrigation Investigation*. Soils Bulletin Draft, Rome, FAO.

FAO/Conservation Foundation. 1954. "Soil Erosion, Survey of Latin America," *Journal of Soil and Water Conservation*. Vol. 9 pp. 158-168, 214-237, 275-280.

FAO/UNDP. 1980. "Integrated Watershed Management, Torrent Control and Land Use Development, Nepal, A Reconnaissance Inventory of the Major Ecological Land Units and their Watershed Condition Summary Report," *FAO Technical Report 1*. FAO,SP/NEP24020 Technical Report 1, Rome, FAO/UNDP.

FAO/UNEP/UNESCO/. 1979. *A Provisional Methodology for Soil Degradation Assessments*. Rome, FAO. with 6 maps.

FAO/UNEP. 1984. *Map of Desertification Hazards, Explanatory Note*. Rome, FAO, May.

FAO/UNESCO. *Soil Map of the World*. Prepared by the Food and Agricultural Organization of United Nations and the United Nations Economics Social and Cultural Organization. 10 Volumes, Paris, FAO. 1978-80.

FAO/UNFPA/IIASA. 1984. *Potential Population Supporting Capacities of Lands in the Developing World*. Rome, FAO.

FAO. 1965. "Soil Erosion by Water Some Measures for its Control on Cultivated Lands," *FAO Land and Water Development Series No. 7/FAO Agricultural Development Paper No. 81*. Rome, FAO.

FAO. 1967. *Pilot Project 1. Watershed Management for the Natural Shikma, Israel, Vol. II Physical Resources*. FAO/SF, 6/ISR Rome, UNDP/FAO.

FAO. 1969. *Soil Survey and land Classification Required for Feasibility Studies of Water Development Projects*. FAO Regional Commission on Land and Water Use in the Near East LA, LWV/69/4 Rome, FAO, June 9.

FAO. 1970. *Colombia Estudio de Preinversion para el Desarrollo Forestal de los Valles del Magdalena y del Sinu Informe sobre los resultados de Proyecto Conclusiones y Recommendaciones*. FAO,SF/COL 14 Terminal Report. Rome, FAO, December.

FAO. 1971. "Invertariacion y Demostraciones Forestales Panama Rehabilitacion de Cuercas Hidrograficas Rios Chioriqui Viejo y Caldera," *FAO Technical Report 6*. FAO, SF/PAN 6. Rome, FAO.

FAO. 1973a. *Guides for Planning Forestland Rehabilitation Work*. FAO/ROK/67/523 Project Report V Seoul, FAO Republic of Korea Ministry of Home Affairs, Office of Forestry.

FAO. 1973b. *Upland Development and Watershed Management, Korea, Community Development and its Place in Comprehensive Watershed Management*. AGL/ROK 67/522 Working Paper 12, Seoul, FAO, October.

FAO. 1973c. *Upland Development and Watershed Management, Korea, Land Conversion Berch Terraces and Soil Conservation on Uplands*. AGL/ROK 67/522, Working Paper 3, Seoul, FAO, August.

FAO. 1973d. *Upland Development and Watershed Management Korean Methodology for Economic Evaluation of Comprehensive Watershed Management*. AGR/ROK/67/522 Working Paper 1. Seoul, FAO, October.

FAO. 1974. "Shifting Cultivation and Soil Conservation in Africa," *FAO Soils Bulletin 24*. Papers presented at the FAO/SIDA/ARCN Regional Seminar. Rome, FAO.

FAO. 1976a. *Upper Silo Watershed Management and Upland Development, Indonesia, Termination Report*. AG,DP/INS/72/001 Rome, FAO/UNDP.

FAO. 1976b. "Soil Conservation and Management in Developing Countries Report of an Export Consultation Held in Rome 22-26 November," *FAO. Soils Bulletin 33*. Rome, FAO.

FAO. 1976c. "Some Aspects of Watershed Management Economics," Compiled and Edited by M. J. Gauchon, UNDP/SF/LAS/72/006, Field Termination Document No. 7, Solo, Indonesia, FAO, August.

FAO. 1977a. "Guidelines for Watershed Management", *FAO Conservation Guide No. 1*. Rome, FAO.

FAO. 1977b. Assessing Soil Degradation Report of an FAO/UNEP Export Consultation held in Rome 18-22 January. *FAO Soils Bulletin 34*. Rome, FAO/UNEP.

FAO. 1977c. "Guidelines for Watershed Management," *FAO. Conservation Guide No. 1*. Rome, FAO.

FAO. 1977d. *Guidelines for the Development of Less Favourable Environment Areas, A Comprehensive Integrated Watershed Development Approach*. AGS/MISC/77/2 March.

FAO. 1978a. Lutte Contre L'Erosion et Conservation des Sols Marue Programmation et Realisation des Travaux dans Les Bassins Versont des Montagnes du Nord du Maroc. Report to the Government of Morocco. FAO Technical Report 3. AG, DP/MOR/71/536, Rome, FAO.

FAO. 1978b. *Report on the Agro-Ecological Zones Project*. Volume I, *Methodology and Results in Africa*. Rome, FAO.

FAO. 1978c. *Agricultural Research in Developing Countries volume 1 - Research Institutions, Current Agricultural Research Information System*. Rome, FAO.

FAO. 1979d. "Land Evaluation Criteria for Irrigation Report of Ian Export Consultation," Rome. February 27-March 2 *World Soil Resources Reports 50*. Rome, FAO.

FAO. 1979e. *Report on the Regional Study on "The Present Situation and Potential Hazards of Soil Degradation in Ten Countries of the N. E. Region,"* Sixth Session of the Regional Commission on Land and Water Use in the Near East, AGL/RNEA,LWV/79/7 February.

FAO. 1979f. "Rural Development in the Khomokhoana and Adjacent Areas, Lesotho, Final Soil Conservation Report," *Technical Document 56*. GCP/LES/009SWE Leribe, FAO, November.

FAO. 1980a. *Sixteenth FAO Regional Conference for Latin America Soil Conservation as a Means of Increasing Food Production in Latin America*. LARC/80/3 Rome, FAO, May.

FAO. 1980b. *Report on the Study Tour FAO/SID/CIDIAT on Incentives for Community Involvement in Forestry and Conservation Programmes.* GCP/INT/347/SWE Rome, FAO.

FAO. 1980c. *Report on the Second FAO/UNFPA Expert Consultation on Land Resources for Populations of the Future*. Rome, FAO.

FAO. 1981a. *Soil and Water Conservation*. Committee on Agriculture, Sixth Session, COAG/81/8 Rome, March 25-April 3.

FAO. 1981b. *Review of Field Programmes*. 1980-1981. C81/4 September Rome, FAO.

FAO. 1981c. Investment Centre. *Introduction Paper on Socio-Political Aspects*. Paper prepared for Informal Workshop on the Economic Aspects of Land Conservation and Humid and Semi-Humid Tropical Conditions, June 30-July 3, Rome.

FAO. 1981d. *Fourth Meeting of the West African Sub-Committee for Soil Correlation and Land Evaluation*. World Soil Resources Reports, No. 53, Rome, FAO.

FAO. 1981e. *Notes on the Preliminary Results of the Social Aspects Survey*. JAM/78/006, GCP/JAM-005 NOR, Extension Section Kingston, Government of Jamaica/UNDP/FAO, June.

FAO. 1981f. *Soil and Water Conservation*. COAG/81/Draft Report III Rome, FAO, April.

FAO. 1981g. *Torrent Control Terminology*. FAO Conservation Guide 6, Rome, FAO.

FAO. 1981h. *Report of an Informal Workshop on Economic Aspects of Land Conservation under Humid Tropical Conditions*. FAO, Rome June 30-July 3.

FAO. 1981i. *Production Yearbook*. Rome, FAO.

FAO. 1982a. *Land Evaluation for Forest Resources Use*. FAO Forestry Department, October.

FAO. 1982b. *World Soil Charter*. Rome, FAO.

FAO. 1982c. *Amenagement des Bassin Versants Maroc Conclusions et Recommendations du Projet*. Report to the Government of Morocco. AG,DP. MOR/78/015. Terminal Report. FAO, Rome.

FAO. 1982d. *Strengthening the National Soil Conservation Programme for Integrated Watershed Development, Jamaica, Terminal Report*. AG,DP/JAM78/006 and GCP/JAM-005/NOR Rome, UNDP/FAO.

FAO. 1982e. *Proceedings of the Government Consultation on Watershed Management for Asia and the Pacific*. FAO,RAS/81/053 Katmandu, Nepal, 5-13 December Rome, FAO.

FAO. 1983a. *Suelos, Climas, Alimentalion Y Poblacion*. Rome, FAO.

FAO. 1983b. "Guidelines on Land Evaluation and Land Classification for Irrigated Agriculture," Draft *FAO Soils Bulletin*. Rome, FAO.

FAO. 1984. *Protect and Produce, Soil Conservation In Development*. Rome; FAO.

Farvar, M.T, and J.P. Milton eds. 1972. *The Careless Technology Ecology and International Development.* Garden City, New York, The Natural History Press.

Fearnside, P.M. 1979a. *The Simulation of Carrying Capacity for Human Agricultural Populations in the Humid Tropics.* Manaus, Brazil, Instituto Nacional des Pesquisas da Amazonia.

Fearnside, P.M. 1979b. "Cattle Yield Prediction for the Transamazon Highway of Brazil," *Interciencia.* Vol. 4, No. 4 July-August, pp. 220-226.

Fearnside, P.M. 1979c. "The Development of the Amazon Rainforest, Priority Problems for the Formulation of Guidelines," *Interciencia.* Vol. 4, No. 4 July-August, pp. 220-226.

Fearnside, P.M. and J.M. Raskin. 1980. "Jari and Development in the Brazilian Amazon," *Interciencia.* Vol. 5, No. 3 May-June, pp. 146-156.

Feder, G, R. Just, and D. Zilberman. 1981. "Adoption of Agricultural Innovations in Developing Countries, A Survey," *World Bank Staff Working Paper No. 444.* February.

Ferejohn, J, and T. Page. 1978. "On the Foundation of Intertemporal Choice," *American Journal of Agricultural Economics.* Vol. 60, No. 2 May pp. 269-275.

Fiddles, D. 1980. "The Effect of Land Use on Flood Flow Peaks from Small East African Basins", in *Casebook of Methods of Computation of Quantitative Changes in the Hydrological Regime of River Basins Due to Human Activities.* UNESCO.

Fields, G.S. 1975. "Higher Education and Income Distribution in a Less Developed Country," *Oxford Economic Papers.* Vol. 27, No. 2 July pp. 245-259.

Fields, S. 1979. *Where Have the Farm Lands Gone*? National Agricultural Lands Study. Washington, D.C.

Finsterbusch, K, and C.P. Wolf, eds. 1977. *Methodology of Social Impact Assessment.* Stroudburg Pennsylvania, Dowder, Hutchinson and Ross.

Fischer, D.W. 1990. Public Policy Aspects of Beach Erosion Control, The Public Interest Requires That All Relevant Interests Have Access to Decision-Making *American Jrnl of Economics & Sociology.* Vol. 49 2 Apr pp. 185-197.

Fisher, A. 1981. *Resource and Environmental Economics.* Cambridge, Cambridge University Press.

Fitz-James, B.P. 1983. "Channela nd Reservoir Sedimentation," Honolulu, East-West Center, Environment and Policy Institute.

Flannery, R.D. 1974a. *Quantifying Land Damage in Latin America Caused by Wind Erosion, A New Approach to the Management and Conservation of Soil and Water Resources.* Report to the Organization of America States, Washington D.C.

Flannery, R.D. 1974b. *The Ecological Undermining of Food-Producing Systems in Latin America Caused by Soil Erosions.* Report to the Organization of American States, Washington, D.C.

Flannery, R.D. 1974c. *Erosion, An International Menace -The Guatemalan Situation*. Report of the Organization for American States, Washington D.C.
Flannery, R.D. 1980. *Summary of Latin American Erosion Control*. Report to the Organization of American States, Washington D.C.
Flemming, W.M. 1979. *Environmental and Economic Impacts of Watershed Conservation on a Maps Reservoir Project in Ecuador*. Paper presented at East-West Center Conference on Extended Benefit Cost Analysis.
Flemming, W.M. 1983. "Phewa Tal Catchment Management Program, Benefits and Costs of Forestry and Soil Conservation in Nepal," in *Forest and Waterside Development and Conservation in Asian and the Pacific*. Hamilton ed. pp. 217-288.
Floyd, B. 1965. "Soil Erosion and Deterioration in Eastern Nigeria, A Geographic Appraisal," *Nigeria Geographic Journal*. Vol. 8, pp. 33-44.
Foster, G.R. and R.E. Highfall. 1983. "Effects of Terraces on Soil Loss, USLE P Factor Values for Terraces," *Journal of Soil and Water Conservation*. Vol. 38, No. 1 January-February. pp. 48-51.
Fournier, F. 1960. *Climat et erosion*. Paris, Presses Universitaires de France.
Fournier, F. 1967. "Research on Soil Erosion and Soil Conservation in Africa," *African Soils*. Vol. 12, No. 1 January-April pp. 53-94.
Fournier, F. 1979. "Wuestionale Methode, Comment approacher un bassins versarte en vue de l'amenagement?," *Hommes, Terre et Eauz*. Vol. 9, No. 30 January-February, pp. 21-26.
Fox, K.A. 1989. Agricultural Economists in the Econometric Revolution, Institutional Background, Literature and Leading Figures *Oxford Economic Papers UK*. Vol. 41 1 Jan pp. 53-70.
Framji, K.K. and B.C. Garg. 1976. "Flood Control in the World, A Global View," *International Commission on Irrigation and Drainage*. Volumes I and II. New Delhi, India.
Freeman, A.M. 1977. "Equity, Efficiency and Discounting, The Reasons for discounting intergnerational effects," *Futures*. Vo. 9, No. 5 October, pp. 375-376.
Freeman, A.M. 1979. *The Benefits of Environmental Improvement, Theory and Practice*. Baltimore, The Johns Hopkins University Press.
Freeman, M. 1983. "Nepal's Environmental Struggle," *Horizons*. U.S. Agency for International Development. February, pp. 9-10.
Frenette, D. 1982. *et al.* "Modelisation de l'Alluvionnement de la Retenue de Peligre, Haiti," Paper presented to the 14th Congress of Large Dams, Rio de Janeiro. Proceedings pp. 93-110.
Friedrich, K.H. 1977. "Farm Management Data Collection and Analysis An Electronic Data Processing, Storage and Retrieval Systems," *FAO Agricultural Services Bulletin 34*. Rome, FAO.
Frye, W.W., S.A. Ebelhar, L.W. Murdock, and R.L. Blevins. 1982. "Soil Erosion Effects on Properties and Productivity of Two Kentucky Soils," *Soil Science Society of American Journal*. Vol. 46, No. 5 pp. 1051-1055.

Frye, W.W., O.L. Bennett, and G.J. Buntley. 1983. *Restoration of Crop Productivity on Eroded or Degraded Soils.* " Soil Erosion and Crop Productivity Symposium ASA/CSSA/SSSA, March 1-3, Denver, Colorado.

Furon, R. 1967. *The Problem of Water, A World Study*. Translated by P. Barnes, New York, American Elsevier.

Gaffney, M.M. 1961. "Land and Rent in Welfare Economics," Acerman, J. and et al. eds, *Land Economics Research.* Resources for the Future. John Hopkins Press. Baltimore, USA.

Gardner, B.D. 1977. "The Economics of Agricultural Land Preservations," *American Journal of Agricultural Economics.* Vol. 59, No. 5 December Proceedings, Issue pp. 1028-1035.

Gass, S. 1981. *Decision-Aiding Models - Validation, Assessment and Related Issues*. Working paper MS/S 81-022, Dept. of Management and Statistics, University of Maryland.

Gauchon, M.J. 1980. *Strengthening the National Soil Conservation Programme for Integrated Watershed Management.* AG,DP/JAM/78/006, Consultant's Report, W. P. 8/Econ, Kingston, FAO, December.

Genaadijer, A.N. 1981. "General Soil Maps of the United States," translated from Bassin *Pochvovedeniye.* No. 7, pp18-26 *Soviet Soil Science.* July-August, No. 4.

George, P. and G. King. 1971. "Consumer Demand for Food Commodities in the United States with Projections for 1980," *Giannini Foundation Monograph.* Number 26, March.

Ghabbour, S.I. 1972. "Some Aspects of Conservation in Sudan," *Biological Conservation.* Vol. 4, No. 3, April.

Gil, R. 1979. *Watershed Development with Special Reference to Soil and Water Conservation.* FAO Soils Bulletin 44. Rome, FAO.

Giordano, E. and Puglisi, S. 1979. "Rapport National Italien," *Hommes, Terre et Eaux.* Vol. 9, No. 30 January-February, pp. 153-158.

Gittenger, J. 1982. *Economic Analysis of Agriculture Projects.* Baltimore, The Johns Hopkins University Press, 2nd edition for World Bank.

Glantz, M.H. and M.E. Krentz. 1981. "Are Solutions to Desertification in the West African Sahel Known but not Applied?," *Desertification Control.* No. 5, December pp. 9-12.

Glesinger, E. 1945. "Forest Production in a World Economy," *American Economic Review.* Vol. 35, No. 2 May pp. 120-129.

Glick, P, "Soil Conservation, Highlights of Political and Legal Arrangements," in *Soil Conservation Policies in an Assessment.* pp. 18-24.

Glymph, L.M. and H.N. Holtman. 1969. "Land Treatment in Agricultural Watershed Hydrology Research," in *Effects of Watershed Changes on Streamflows.* Austin, University of Texas Press, pp. 44-68.

Glymph, L.M. 1975. "Evolving Emphasis in Sediment-Yield Predictions," ERS/USDA *Present and Prospective Technology for Predicting Sediment Yields and Sources.* ARA-S-40, V.S.A, pp. 1-4.

Goldberg, E.D. and J.J. Griffin. 1964. "Sedimentation Rates and Mineralogy," *South Atlantic Journal of Geophysical Research*. Vol. 69, No. 20 October 15, pp. 4293-4309.

Goldberg, J. 1980. *Indicative Economics of Soil Conservation Works.* " Memorandum. Washington, D.C. World Bank, December 22.

Goldman, M. 1979. *The Spoils of Progress, Environmental Pollution in the Soviet Union*. Cambridge, Mass. The MIT Press.

Golladay, F. and M. Burns. 1983. *Economic Evaluation of Flood Control Activities*. Draft Report Washington, D.C. World Bank, January 25.

Gonzalez, L.A. 1979. *Un Modelo de Evaluacion Economica Para Provectos de Monejo de Cuevas*. M.S. Thesis Universidad de Merida, Venezuela, Universidad de los Andes.

Gould, B.W., Saupe, W.E. and Klemme, R.M. 1989. "Conservation Tillage, The Role of Farm and Operator Characteristics and the Perception of Soil Erosion," *Land Economics*. Vol. 65 2 May pp. 167-182.

Graf, H. 1973. "Bodenerosion und Schutzmassnahmen in marokkanischen Rifgebirge," oil Erosion and Measures of Control in the Moroccan Rif Range. *Schweizerische Zeitschrift Fur Forstwesen*. Vl. 124, No. 11, pp. 863-870.

Grant, K.E. 1975. "Erosion in 1973-74, The Record and The Challenge," *Journal of Soil and Water Conservation*. Vol. 30, No. 1, January-February.

Greenland, D.J. and R. Lal. 1977. *Soil Conservation and Management in the Humid Tropics*. New York, John Wiley and Sons.

Griliches, Z. 1958. "Research Costs and Social Returns, Hybrid Corn and Related Innovations," *Journal of Political Economy*. Vol. 66, No. 5 October pp. 419-432.

Guntermann, K.L. Ming T. Lee, and E.R. Swanson. 1976. "The Economics of Off- Site Erosion," *The Annal of Regional Science*. August pp. 117-26.

Gupta, G.P. 1975. "Sediment Production - Status Report on Date Collection and Utilization," *Soil Conservation Digest*. Vol. 3, No. 2 September pp. 10-21.

Gupta, S., R. Prasad, and R. Pandey. 1973. "An Economic Evaluation of Soil Conservation Measures in Varanasi District, U. P," *Indian Journal of Agricultural Economics*. Vol. 28, pp. 205-211.

Gupta, S.K. and Ram Babu. 1977. "Studies of Efficiency of Contour Farming, Channel Terracing and Channel Terracing with Graded Furrows for Erosion Control on 4% Sloping Cultivation Land," *Soil Conservation Digest*. Vol. 5(2) October, pp. 29-32.

Hadley, R.D. and L.M. Shown. 1976. *Relation of Erosion to Sediment Yield*. in *Proceedings of the Third Federal Inter-Agency Sedimentation Conference*. March 22-25, Denver, Colorado, pp. 1-132 - 1-139.

Hagan, R.M., H.R. Haise, and T.W. Edminster. 1967. "Irrigation of Agricultural Lands," *Agronomy Series, No. 110*. Madison, Wisconsin, American Society of Agronomy.

Hager, L.L. and P.T. Dyke. 1979. *The RCA Yield/Soil Loss simulator, Methods, Data and Results for Selected Crops*. Draft USDA/ESLS/NRED, November.

Hager, L.L. and P.T. Dyke. 1980. *Yield/Soil Loss Relationships*. Paper prepared for Workshop or Influence of Erosion on Soil Productivity. February 26-28.

Hahn, C. 1982. *The Economic Rationale for Protection and Management of Natural Areas in Developing Countries*. Draft. N. P, Natural Resources Defense Council, April.

Halcrow, H., E.O. Heady and M. Cotner, eds. 1982. *Soil Conservation Policies Institutions and Incentives*. Ankeny, Iowa, Soil Conservation Society of America.

Hall, A.J. 1981. "Flash Flood Forecasting," *WMD Operational Hydrology Report*. No. 18, Geneva.

Hamilton, J. and C. Pongtanakorn. 1982. *The Economic Impact of Irrigation Development in Idaho, An Application of Marginal Input-Output Methods*. Moscow, Idaho, University of Idaho, College of Agriculture. Presented at American Agricultural Economics Meetings, Logan, Utah.

Hamilton, L.S. ed. 1983. *Forest and Watershed Development and Conservation in Asia and the Pacific*. Boulder, Colorado, Westview Press.

Hamilton, L.S. 1985. "Towards Clarifying the Appropriate Mandate in Forestry Rehabilitation and Management," Paper prepared for Expert Consultation of Strategies, Approaches and Systems for Integrated Watershed Management. Katmandu, March.

Hamilton, L.S. and P.N. King. 1983. *Tropical Forested Watersheds, Hydrologic and Soils Response to Major Use or Conversions*. Boulder, CO, Westview Press.

Hamilton, L.S. and S.C. Snedaker, eds. 1984. *Handbook for Mangrove Area Management*. Honolulu, East-West Center.

Hammer, J.S. 1986. 'Subsistence First', Farm Allocation Decisions in Senegal *Jrnl of Development Economics Netherlands*. Vol. 23 2 Oct pp. 355-369.

Han, Q.W. and Z.J. Tong. 1982. "The Impact of Danjiangkou Reservoir on the Downstream River channel and the Environment," Paper presented to the 14th Congress of Large Dams, Rio de Janeiro. Proceedings, pp. 189-200.

Harberger, A. 1971. "Three Basic Postulates for Applied Welfare Economics, An Interpretive Essay," *Journal of Economic Literature*. Vol. 9, pp. 785-98.

Harou, P.A. and J.G. Massey. 1982. "Monitoring Forestry Projection, The Alternatives Test," *Agricultural Administration*. Vol. 9 pp. 139-146.

Harris, J.M. 1991. "World Agricultural Production, Growth Paths and Environmental Effects," *American Economist*. Vol. 35 1 Spring pp. 62-74.

Harrison, P. 1983. "Land and People, the Growing Pressure," *Earthwater*. November 13, pp. 1-8.

Harrold, L.L. and W.M. Edwards. 1972. "A Severe Rainstorm Test of Nontill Corn," *Journal of Soil and Water Conservation*. Vol. 27, No. 1 January-February.

Hartwick, J.M. 1977. "Intergenerational Equity and the Investing of Rests from Exhaustible Resources," *American Economic Review*. Vol. 67, No. 5 December pp. 972-974.

Hartwick, J.M. 1978a. "Substitution Among Exhaustible Resources and Intergenerational Equity," *Review of Economic Studies*. Vol. 45, No. 2 June, pp. 347-54.

Hartwick, J.M. 1978b. "Investing Returns from Depleting Renewable Resource Stocks and Intergenerational Equity," *Economic Letters*. Vol. 1 pp. 85-88.

Hartwick, J.M. 1989. "On the Development of the Theory of Land Rent," *Land Economics*. Vol. 65 4 Nov pp. 410-412.

Harza Engineering Company. 1980. *Environmental Design Considerations for Rural Development Projects*. U.S. Agency for International Development, Washington, D.C. October.

Haveman, R.H. 1977. "The Economic Evaluation of Long-Run Uncertainties," *Futures*. Vol. 9, No. 5 October, pp. 365-374.

Haveman, R.H. and J. Krutilla. 1968. *Unemployment, Idle Capacity and the Evaluation of Public Expenditures*. Baltimore, Johns Hopkins University Press.

Hayami, Y. and V. Ruttan. 1971. *Agricultural Development, An International Perspective*. Baltimore, The Johns Hopkins University Press.

Hayami, Y. and W. Peterson. 1972. "Social Returns to Public Information Services, Statistical Reporting of U.S. Farm Commodities," *American Economic Review*. Vol. 62, pp. 119-30.

Hayman, E. 1981. "The Valuation of Extramarket Benefits and Cuts in Environmental Impact Assessment," *Environmental Impact Assessment Review*. Vol. 2, No. 3.

Heady, E.O., K.J. Nicol, and J.C. Wade. 1976. "Economic Trade-offs to Limit Nonpoint Sources of Agricultural Pollution," *Water, Air and Soil Pollution*. Vol. 5 pp. 415-430.

Heady, E.O. and G.F. Vocke. 1978. "Trade-offs Between Erosion and Production Costs in U.S. Agriculture," *Journal of Soil and Water Conservation*. Vol. 33, No. 5 September-October pp. 227-30.

Heady, E.O. and D.R. Daines Jr. 1982. "Short-term and Long-term Implications of Soil Loss Control on U.S. Agriculture," *Journal of Soil and Water Conservation*. Vol. 37, No. 2 March-April pp. 109-113.

Hedfors, L. 1981. *Evaluation and Economic Appraisal of Soil Conservation in a Pilot Area*. Nairobi, Soil and Water Conservation Branch, Ministry of Agriculture, August.

Hefny, K. 1979. "Rapport du comite national MB d'Egypte sur le controle et l'Amenagement du Nil," *Homme, Terre et Eaux*. Vol. 9, No. 30 January-February, pp. 169-175.

Heft, F.E. "Political, Social and Economic Aspects of Soil Erosion and Sedimentation Control Programs," in *Soil Erosion and Sedimentation*. American Society of Agricultural Engineers, pp. 23-30.

Henderson, A. 1941. "Consumer's Surplus and the Compensating Variation," *Review of Economic Studies*. Vol. 8, February.

Henderson, J. and R. Quandt. 1971. *Microeconomic Theory*. New York, McGraw- Hill, Inc.

Henneberry, D.M. and Barrows, R.L. 1990. "Capitalization of Exclusive Agricultural Zoning into Farmland Prices," *Land Economics*. Vol. 66 3 Aug pp. 249-258.

Henry, C. 1974. "Option Values in the Economics of Irreplaceable Assets," *Review of Economic Studies*. pp. 89-104.

Hicks, J. 1939. "The Foundations of Welfare Economics," *Economic Journal*. Vol. 49, pp. 696-712.

Hicks, J. 1941. "The Rehabilitation of Consumers' Surplus," *Review of Economic Studies*. Vol. 8, February.

Hicks, J. 1942. "Consumers' Surplus and Index Numbers," *Review of Economic Studies*. Vol. 9, Summer.

Hitchens, M.T., D.J. Thampapillai, and J.A. Sinden. 1978. "The Opportunity Cost Criterion for Land Allocation," *Review of Marketing and Agricultural Economics*. Vol. 46, No. 3 December pp. 275-293.

Hitzhusen, F. *et al.* 1985. "Private and Social Cost Benefit Perspectives and a Case Application on Reservoir Sedimentation Management," *Water International*. Vol. 10, No. 1 March.

Holeman, J.N. 1968. "The Sediment Yield of Major Rivers of the World," *Water Resources Research*. Vol. 4, No. 4 August, pp. 737-47.

Holly, M. 1980. *Erosion and Environment*. New York, Permagon Press.

Holmes, T.P. 1988. "The Offsite Impact of Soil Erosion on the Water Treatment Industry," *Land Economics*. Vol. 64 4 Nov pp. 356-366.

Homan, A.G, and B. Waybor. 1960. *A Study of Procedure in Estimating Flood Damage to Residential, Commercial and Industrial Properties in California*. SRI Project Nos. 1-2541 and 1-2880 prepared for The United States Soil Conservation Service and the California State Division of Soil Conservation, Marlo Park, California, Stanford Research Institute, January.

Hotelling, H. 1938. "The General Welfare in Relation to Problems of Taxation and of Railway and Utility Rates," *Econometrica*. Vol. 6, pp. 242-26?.

Howard, P.M. 1981. "Impressions of Soil and Water Conservation in China," *Journal of Soil and Water Conservation*. Vol. 36, No. 3 May-June pp. 122-124.

Howe, C. 1979. *Natural Resource Economics*. New York, John Wiley & Sons.

Hudson, N. 1981. *Soil conservation.* 2nd Ed Cornell UP, Ithaca NY.

Hufschmidt, M. *et al.* 1981. *Benefit-cost analysis of natural systems and environmental quality aspects of development.* East-West Environment and Policy Institute. Honolulu, Hawaii.

Hufschmidt, M. 1983. *Environment, natural systems and development, an economic valuation guide.* East-West Environment and Policy Institute. Honolulu, Hawaii.

INDERENA. 1982a. *El Incentiuo en la Conservacion de Aguas v Suelos*. Bogota, Ministry of Agriculture, November.

INDERENA. 1982b. *Programa de Investigacion en Microcuencas*. Bogota, Ministry of Agriculture, September.

INDERENA. 1983. *Cuencas Prioritarias para Ordenamiento y Menajo*. Draft Report. Bogota, February.

India, Government of. 1976. *Report of the National Commission on Agriculture 1976*. par V Resource Development. New Delhi, Ministry of Agriculture and Irrigation.

India, Government of 1978. *Report of the Working Group on Integrated Action Plan for Flood Control In Indo-Gangetic Basin*. Ministry of Agriculture and Irrigation Department of Agriculture, New Delhi, December.

India, Government of, Ministry of Irrigation and Power. 1972. *Fifth Plan Position Paper on Flood Control, Drainage and Anti-Waterlogging, Anti Sea Erosion*. New Delhi May.

India, Government of. 1981. *Report of the Sub-Group on Transport Planning for the Himalayan Region*. Planning Commission, December.

Institute of Ecology. 1981. "An Environmental Profile of Guatemala, Assessment of Environmental Problems and Short and Long-term Strategies for Problem Solution," report prepared for Agency for International Development, U.S. Department of State, Athens, Georgia, University of Georgia, Institution of Ecology, May.

Instituto Nacional de los Recursos Naturales Renovables y del Ambiante INDERENA. 1983a. National Institute for Renewable Natural Resources and the Environment *Plan de Investigacion en Microcuencas*. Bogota, Ministry of Agriculture, June.

Instituto Nacional de los Recursos Naturales Renovables y del Ambienta INDERENA. 1983b. *Plan De Investigacion Agrosilvopastoril*. Bogota, Ministry of Agriculture, June.

Inter-American Development Bank IDB. 1982a. *Sector Forestal, Criterios Especiales Para Evaluacion Socioeconomica de Proyectos*. Washington, D.C. March.

Inter-American Development Bank IDB. 1982b. *The Inter-American Development Bank in the Forestry Sector in Latin America*. Washington, D.C. Fishery and Forestry Development Section, IDB, June.

International Commission on Large Dams. 1938. *Transactions Second Congress on Land Dams*. Vol. V. Washington, U.S. Gov. Printing Office.

International Commission on Large Dams. 1951. *Fourth Congress on Large Dams Transactions*. Vol. I-IV New Delhi, World Power Conference January.

International Land Development Consultants. 1981. *Agricultural Compendium for Rural Development in the Tropics*. Amsterdam, Elsevier Scientific Publishing Company.

Ivan, J.C. 1977. "Data Requirements for the Optimization of Reservoir Design and Operating Rule Determination," UNESCO/WMO/LAHS, pp. 335-347.

Jades, G.V. and Whyte, R.O. 1939. *The Rape of the Earth - A World Survey of Soil Erosion*. London, Faber.

Jakeman, A., D.G. Day and A.K. Dragun. 1985. *Environmental Quality Control*. CRES Monograph Series No. 11, with. ANU, Canberra.

Jakobsson. K.M and A.K. Dragun. 1991. "Water and soil management in New Zealand", *Journal of Environmental Management*. Vol33,1-16.

James, I.C. "Data Requirements for the Optimization of Reservoir Design and Operating Rule Determination," in UNESCO/WMO/LAHS, pp. 335-347.

Jansen, J.M.L. and R.B. Painter. 1974. "Predicting Sediment Yield from Climate and Topography," *Journal of Hydrology*. Vol. 21, pp. 371-380.

Jansma, J.D. 1964. *Secondary Effects of Upstream Watershed Development, Roger Mills Country Oklahoma*. PhD Dissertation. Ap, Oklahoma State University, May.

Jarocki, W. 1963. *A Study of Sediment*. Translated from Polish. N. S. F.

Jenny, H. 1980. *The Soil Resource Origin and Behaviour*. New York, Spinger-Verlag.

Jensen, H.R., E.O. Heady, and R.V. Baumann. 1955. "Costs, Returns and Capital Requirements for Soil-Conserving Farming or Rented Farms in Western Iowa," *Research Bulletin 423*. Agricultural Experiment Station, Iowa State College. Ames, Iowa, March.

Jevons, W.S. 1965. *The Theory of Political Economy*. New York, Augustus M. Kelley.

Johannsen, C.J. and T.W. Barney. 1981. "Remote Sensing Applications for Resource Management," *Journal of Soil and Water Conservation*. Vol. 36, No. 3, May-June, pp. 128-124.

John, B.C. and W. van de Goot. 1976. *Soil Conservation Measures in the Upper Solo Basin*. FAO report IAS/72/006, Termination Field Document, Soil Conservation, Solo, Indonesia, FAO, June.

Johnson, G.V. 1981. "Soil Fertility, An Analogy," *Crops and Soil Magazine*. December, pp. 8-11.

Johnson, S. and G. Rausser. 1977. "Systems Analysis and Simulation, A Survey of Applications in Agricultural and Resource Economics," *A Survey of Agricultural Economics Literature*. Vol. 2, L. Martin ed. Minneapolis, University of Minnesota Press.

Johnson, S.H. and S. Kolavalli. 1984. "Physical and Economic Impacts of Sedimentation on Fishing Activities," Nam Pong Basin, Northeast Thailand," *Water International*. 9 pp. 185-188.

Johnston, B.F. and J.W. Mellor. 1961. "The Role of Agriculture in Economic Development," *American Economic Review*. Vol. 51, No. 4, September, pp. 566-593.

Jones, P.M.S. 1978. "Intergenerational Equity, A Response," *Futures*. February, pp. 68-70.

Jose, D.H. and D.D. Markland. 1981. "Land Value in Two Soil Zones with and without Fertilizers, An Economic Analysis," *Journal of Soil and Water Conservation*. Vol. 36, No. 2, March-April, pp. 114-116.

Joshua, W.D. 1977. "Soil Erosive Power of Rainfall in the Different Climate Zones of Sri Lanka", in UNESCO and IAHS, *Erosion and Sediment Transport in Island Water*. pp. 51-61.

Joyce, C. 1981. "Peru's Ecological Disaster," *New Scientist*. June 18, p. 746.

Judson, S. 1968. "Erosion of the Land or What's Happening to our Continents?," *American Scientist*. Vol. 56, No. 4, pp. 356-374.

Jurgensmeyer, J.C. and J.B. Wadley. 1974 "The common lands concept , A "commons," solution to a common environmental problem," *Natural Resources Journal*. 14 , 368-81.

Just, R. and D. Hueth. 1979. "Welfare Measures in a Multimarket Framework," *American Economic Review*. Vol. 69, pp. 947-54.

Just, R., D. Hueth and A. Schmitz. 1982. *Applied Welfare Economics and Public Policy*. Englewood Cliffs, N. J, Prentice-Hall, Inc.

Just, R.E. and D.A. Zilberman. 1988. "Methodology for Evaluating Equity Implications of Environmental Policy Decisions in Agriculture," *Land Economics*. Vol. 64 1 Feb pp. 37-52.

Kaldor, N. 1939. "Welfare Propositions of Economics and Interpersonal Comparisons of Utility," *The Economic Journal*. Vol. 49, pp. 549-52.

Kasal, J. 1976. "Trade-offs between Farm Income and Selected Environment Indicators, A Case Study of Soil Loss, Fertilizer and Land Use Constructs," United States Department of Agriculture Economic Research Service, *Technical Bulletin No. 1550*. August.

Kaul, U.N. 1980. *Impact of Forest Land Uses on Environment in India*. Paper presented at the IUFRO/MAB Conference on Research on Multiple - Use of Forest Resources, May 18-23, Flagstaff, Arizona.

Keller, P. and J.P. Bouchard. 1983. *Etude Bibliographique De L'Alluvionnement Des Retenues Par Les Sediments Fins*. International Re Electricite de France, July.

Kelley, H.W. 1983. "Keeping the Land Alive Soil Erosion, Its Causes and Cures," *FAO Soils Bulletin 50*. Rome, FAO.

Kellogg, C.E. and A.C. Orvedal. 1969. "Potentially Arable Soils of the World and Critical Measures for Their Use," *Advances in Agronomy*. Vol. 21 pp. 109-170.

Kersetes, D. and K. Easter. 1981. "A Review and an Annotated Bibliography of Studies of Soil Conservation Programs, Practices and Strategies," St. Paul, University of Minnesota, Dept. of Agricultural and Applied Economics, Staff Paper pp. 81-1.

Kharkwal, S. 1971. Slope Studies on a Himalayan Terrain. *The National Geographic Journal of India*. 17 1, 147-148.

Khosla, A.N. 1953. *Silting of Reservoirs*. Central Board of Irrigation and Power Publication No. 51 Simla, Government of India Press.

King, D.A. and J.A. Sinden. 1988. "Influence of Soil Conservation on Farm Land Values", *Land Economics*. Vol. 64 3 Aug pp. 242-255.

Kingu, P.A. 1980. *World Map of Erosivity*. M. S. Thesis n. p, Cranfield Institute of Technology, National College of Agricultural Engineering, September.

Kirby, M. and R. Morgan eds. 1980. *Soil Erosion*. New York, John Wiley Sons.

Klingebiel, A.A. 1967. "Land Resources Available to People in the United States," *Proceedings of the 16th Annual Meeting Agricultural Research Institute, National Research Council*. Washington, D.C. Reprint.

Klingebiel, A.A. 1972. *Soil and Water Management to Control Plant Nutrients in Natural Water*. Paper prepared for the FAO Conference on the Effects of Intensive Fertilizer Use on the Humid Environment, January 25-28, Rome, Italy.

Klingebiel, A.A. 1973. "Nutrient Enrichment of Natural Waters," *Qual. Plant. Mater. Veg*. Vol. 22, No. 3-4, pp. 223-248.

Knisel, W.G. *et al*. 1982. "Nonpoint-Source Pollution Control, A Resource Conservation Perspective," *Journal of Soil and Water Conservation*. Vol. 37, No. 4, July-August, pp. 196-199.

Kock, W. 1982. *Principles and Technologies of Sustainable Agriculture in Tropical Areas, A Preliminary Assessment*. " Draft Consultants Report prepared for the World Bank, September.

Kohnke, H. and A.R. Bertrard. 1959. *Soil Conservation*. New York, McGraw Hill.

Kotok, E.I. 1945. "International Policy on Renewable National Resources," *American Economic Review*. Vol. 35, No. 2 May pp. 110-119.

Krammerer, P.A. and W.G. Batter. 1982. "Sediment Deposition in a Flood Retention Structure after Two Record Floods in Southern Wisconsin," *Journal of Soil and Water Conservation*. Vol. 37, No. 5 September-October pp. 302-304.

Krestovsky, O.L. and S.F. Fedorov. 1970. "Study of Water Balance Elements of Forest and Field Watersheds," in *World Water Balance Proceedings of the Reading Symposium*. Vol. 2, July. See L\IASH/UNESCO/WMO.

Krog, D., S. Bhide, C. Pope and E.O. Heady. 1983. "Effects of Livestock Enterprises on the Economics of Soil and Water Conservation Practices in Iowa," *CARD Report 112*. SWCP Series V. Center for Agricultural and Rural Development, Iowa State University.

Krutilla, J.V. 1967. "Conservation Reconsidered," *American Economic Review*. Vol. 57, No. 4 September pp. 777-786.

Krutilla, J.V. and A. Fisher. 1975. *The Economics of Natural Environments*. Baltimore, The Johns Hopkins University Press.

Kula, E. 1981. "Future Generations and Discounting Rules in Public Sector Investment Appraisal," *Environment and Planning*. Vol. 13, pp. 899-910.

Lal, R. 1976. "Soil Erosion as Alfisols in Western Nigeria, I. Effects of Slope, Crop Rotation and Residue Management," *Geoderma*. Vol. 16 pp. 363-375.

Lamb, F.B. 1987. "The Role of Anthropology in Tropical Forest Ecosystem Resource Management and Development", *Jrnl of Developing Areas*. Vol. 21 4 Jul pp. 429-458.

Lanly, J-P. 1982. "Tropical Forest Resources," *FAO Forestry Paper 30*. Rome, FAO.

Largdale, G.W., J.G. Box Jr., R.A. Leonard, A.P. Barneth, and W.G. Fleming. 1979 "Corn Yield Reduction on Eroded Southern Piedmont Soil," *Journal of Soil and Water Conservation*. Vol. 34, No. 5 September-October pp. 226-228.

Larson, B.A. and M.K. Knudson. 1991. "Whose Price Wins, Institutional and Technical Change in Agriculture," *Land Economics*. Vol. 67 2 May pp. 213-224.

Larson, B.A. and D. Southgate. 1991. "The Causes of Land Degradation Along "Spontaneously," Expanding Agricultural Frontiers in the Third World, Comment; Reply," *Land Economics*. Vol. 67 2 May pp. 260-268.

Larson, D.M. and I.W. Hardie. 1989. "Seller Behavior in Stumpage Markets with Imperfect Information," *Land Economics*. Vol. 65 3 Aug pp. 239-253.

Larson, W.F., F.J. Pierce, and R.H. Dowdy. 1983. "The Threat of Soil Erosion to Long-Term Crop Production," *Science*. Vol. 219 February 4, pp. 498-465.

Leal, J. 1980. *A Framework for the Application of Cost-Benefit Analysis to Environmental Protection Measures*. Report prepared for the Third Intergovernmental Expert Group Machinery on Cost Benefit Evaluation of Environmental Protection Measures, Paris, November 18-20, UNEP/IG. 21/2, 15 October.

Lee, L.K. 1980. "The Impact of Landownership Factors on Soil Conservation," *American Journal Agricultural Economics*. December, pp. 1071-1076.

Lee, L.K. 1981. "Cropland Availability, The Landowner Factor," *Journal of Soil and Water Conservation*. Vol. 36, No. 3 May-June pp. 135-137.

Lee, M.T., A.S. Narayanan, K. Gunterman, and E.R. Swanson. 1974. "Economic Analysis and Sedimentation Hambaugh-Martin Watershed," *Agricultural Economics Research Report 127*. IIEQ Document No. 74-28 Agricultural Experiment Station, University of Illinois, Urbana-Champaign/State of Illinois Institute for Environmental Quality, July.

Leonard, H.J. 1983. *Socio-Economic Aspects of Natural Resource Management*. Summary of remarks prepared for a Workshop on Rural Development Research, U.S. Agency for International Development, October 20-21, n. p. September 30.

LeVeen, E.P. 1981. *Some Economic Considerations for Soil Conservation Policy*. Report prepared by the Public Interest Economics Center for the Natural Resources Defense Council. Washington, D.C. February 9.

Libby, L. 1980. "Who Should Pay for Soil Conservation?," *Journal of Soil and Water Conservation*. Vol. 35, No. 4 July-August pp. 155-156.

Library of Congress. 1978-1982. *Draft Environmental Reports (various countries)*. U.S. Man and the Biosphere Secretariat, Department of State, Washington, D.C.

Liggett, A.J. 1974. "Erosion, Landslides and Sedimentations in Colombia," Proposal to the USAID-Bogota, by Cornell University, Corporacion Informa del Valle del Cauca and Universidad del valle, December.

Lin, S, and S. Ramakrishnan, eds. 1979. "Land Management Issues and Development Strategies in Developing Countries," *Lincoln Institute Monograph No. 78-4*. Cambridge, Mass, Lincoln Institute of Land Policy, March.

Lockeretz, W. 1989. "Secondary Effects on Midwestern Agriculture of Metropolitan Development and Decreases in Farmland," *Land Economics*. Vol. 65 3 Aug pp. 205-216.

Logan, T.S. "Establishing Soil Loss and Sediment Yield Limits for Agricultural Land", *Soil Erosion and Sedimentation*. American Society of Agricultural Engineers, pp. 59-68.

Lowe, M.J. and H.R. Fox. 1982. 'Sedimentation in Tarbele Reservoir," Paper presented to the 14th Congress of Large Dams, Rio de Janiero. Proceedings, pp. 317-340.

Lyles, L., J.D. Dickerson, and N.F. Schmeidler. 1974. "Soil Detachment from Clods by Rainfall, Effects of Wind, Mold Cover and Initial Soil Moisture," *Transactions of the American Society of Agricultural Engineers*. Vol. 17, No. 4, pp. 669-700.

Mac, Chaem, *et al*. 1966. *Review of National Places That Make References to Soil Erosion*. U.S. AID.

Machado, S.A. 1978. "Primera Aproximacion de los Valores del Factor de Erodabilidad en Algunos Suelos Colombianos," *Revista Facultad Nc. de Agronomia*. Vol. 31, No. 1 pp. 1-22.

Maddaruddin, S. 1968. Social and Economic Aspects of Soil Conservation in the Machkund Basin. *Journal of Soil and Water Conservation in India*. 16 1,2, p 85-90.

Magrath, W. 1979. *Influence of Tree Shelterbelts on Agricultural Crop and Livestock Yields*. Memorandum. Washington, D.C. ; World Bank, November 26.

Maler, K.G. 1974. *Environmental Economics, A Theoretical Inquiry*. Baltimore, The Johns Hopkins University Press.

Maletic, J.T. 1952. *Using Soil Survey Information in Land Classification for Irrigation*. Address delivered to the American Society of Agronomy and Soil Science Society of America, November 19, Cincinnati, Ohio.

Maletic, J.T. 1962. *The Relationship Between Land Classification and Engineering in the Planning of an Irrigation Project*. January 11.

Maletic, J.T. 1966. *Land Classification Survey as Related to the Selection of Irrigable Lands*. Paper presented at the Pan American Soil Conservation Congress Section III, Part 3, April, San Paulo, Brazil.

Maletic, J.T. 1967. *Irrigation, a Selective Function - Selection of Project Lands Factors Influencing Selection*. Paper prepared for the International Conference on Water for Peace, May 23-31, Washington, D.C.

Maletic, J.T. 1968. *Sprinkler Irrigation Soils, Climate and Land Classification*. Paper prepared for the 1968 Annual Technical Conference, Sprinkler Irrigation Association, February 27, Denver, Colorado.

Maletic, J.T. 1970. *Land Classification Principles*. Soil Scientist Training Institute, Principles Lecture, August 3.

Maletic, J.T. and M.N. Langley. 1971. *Experience and Trends in Automation of Project and Farm Irrigation Systems*. Paper prepared for the Experts Panel on Irrigation, State of Israel, Ministry of Agriculture, September 6-13, Tel Aviv, Israel.

Maletic, J.T. and W.H. Yarger. 1968. *Engineering and Economic Relationships in the Selection and Classification of Irrigable Lands*. Paper presented at the Fourth Technical Conference on Irrigation, Drainage and Flood Control, March 27-29, Phoenix, Arizona.

Malhotra, A.N. and R.N. Hoon. 1971. "Sedimentation Studies of Bhakra Reservoir," *Irrigation and Power*. Vol. 28, No. 1, January, pp. 37-52.

Malik, M. 1988. "Agriculture, The Wasting of a Continent; Fighting the Arid Land," *Far Eastern Economic Review Hong Kong*. Vol. 141 38 Sep 22, pp. 84-86.

Mann, H.S. 1982. "Revegetation of the Indian Desert", *Desertification Control*. No. 6, April pp. 6-9.

Mann, J. 1977. "Techniques to Measure Social Benefits and Costs in Agriculture, A Survey," *American Economic Review*. Vol. 29, pp. 115-26.

Mannering, J.V. 1983. *et al,. Regional Effects of Soil Erosion on Crop Productivity - Midwest*. Paper prepared for Soil Erosion and Crop Productivity Symposium March 1-3, Denver, Colorado.

Margaris, N.S. 1979. "Le probleme de l'erosion en Grece," *Hommes, Terre et Eaux*. Vol. 9, No. 30 January-February pp. 143-151.

Marshall, A. 1961. *Principles of Economics*. London, MacMillan.

Marshall, A. 1969. "Price Stabilization and Welfare," *Quarterly Journal of Economics*. Vol. 83 May.

Mathur, R.S. 1978. *Methodological Studies for Analysis and Planning of Non-wood Outputs of Forestry*. Study prepared under an FAO Andre Mayer Research Fellowship, June.

Matthus, T.R. 1970. *An Essay on the Principle of Population*. Middlesex, Penguin.

McCarl, B. and G. Nelson. 1983. *Model Validation, An Overview with Some Emphasis on Risk Models*. S-180 Project. San Antonio, Texas.

McCarthy, J.W. *et al,*. 1983. *Somalia Soil and Water Conservation Coal Bardaale and Arabsiyo Project Impact Evaluation*. Draft. Washington, D.C. U.S. Agency for International Development, April.

McConnell, K.E. 1983. "An Economic Model of Soil Conservation," *American Journal of Agricultural Economics*. Vol. 65, No. 1 February, pp. 83-89.

McDonald, H.R. and J.T. Maletic. 1967. *Water Quality Relationships in Irrigation Development*. Paper prepared for the New Mexico Water Conference, March 30-31, Las Cruces, New Mexico.

McGregor, D.F.M. 1980. "An Investigation of Soil Erosion in the Colombian Rainforest Zone," *CATEWA*. Vol. 7, pp. 265-273.

Mclnerney, J. 1976. "The Simple Analytics of Natural Resource Economics," *Journal of Agricultural Economics*. Vol. XXVII, No. 1, January, pp. 31-52.

McMartin, W. 1950. "The Economics of Land Classification for Irrigation," *Journal of Farm Economics*. Vol. 32, No. 4, part 1 November pp. 553-70.

Mejia F.R. 1949. "Compara de Defensa y Restavracion de los Suelos er las Regiones Cafeteras," *Agric Trop. Bogota*. Vol. 5, No. 5 pp. 39-43.

Mendez, R. 1981. "Combatting Desertification in the Sudano-Sahelian Region," *Desertification Control*. No. 5, December pp. 2-8.

Mendoza, R.P. 1981. *Las Cuencas Hidrograficas Su Importancia Actual y Futuro*. Paper prepared for Coloquio Regional Sobre el Agua, June 18-20, Popayan, Colombia. Processed by Instituto Nacional de los Recursos Naturales Renovables y del Ambiente.

Mermel, T.W. 1981. "Major Dams of the World," *Water Power and Dam Construction*. May.

Meta Systems. 1979. *Environmental and Economic Considerations of Watershed Management in Colombia*. Report prepared for the World Bank's Latin American Projects Department December.

Meyer, L.D., W.H. Wischmeier, and G.R. Foster. 1970. "Mulah Rater Required for Erosion Control on Steep Slopes," *Soil Science Society of America*. Vol. 34, No. 6 November-December, pp. 928-31.

Mildner, W.F., J.S. Bali, and H.S. Nandaa, eds. 1972. *National Inventory of Reservoir and Tank Sediment data*. Soil Conservation Division, Central Unit for Soil Conservation, Hydrology and Sedimentation, Government of India, Ministry of Agriculture, New Delhi.

Millington, T. 1982. "Soil Conservation Techniques for the Humid Tropics", *Appropriate Technology*. Vol. 9, No. 2 September pp. 17-18.

Ministry of Agriculture and Cooperatives Thailand. 1980. *Land Reform Areas Development Project*. Draft Report Agriculture. Annex 4, Soil Conservation, Annex 5 prepared by the Foreign Loan Project Office, Bangkok, October 1.

Mishan, E.J. 1948. "Realism and Relevance in Consumer's Surplus", *Review of Economic Studies*. Vol. 15.

Mishan, E.J. 1959. "Communications, Rent as a Measure of Welfare Change", *American Economic Review*. Vol. 49, pp. 386-95.

Mishan, E.J. 1968. "What is Producer's Surplus?," *American Economic Review*. Vol. 58, pp. 1269-82.

Mishan, E.J. 1976. *Cost-Benefit Analysis*. New York, Praeger Special Studies.

Mishan, E.J. 1977a. "The Plain Truth about Consumer Surplus", *Zeitschrift Fur National Okoomie Journal of Economics*. Vol. 1 37, pp. 1-24.

Mishan, E.J. 1977b. "Economic Criteria for Intergenerational Comparisons", *Futures*. Vol. 9, No. 5 October pp. 383-403.

Misra, D.R. Prosad and S. Bhan. 1969. "Impact of Soil Conservation Practices on Crop Production in India,. *Journal of Soil and Water Conservation in India*. 17 1,2, 52-57.

Misra, R. 1974. *Environment aspects of Land-use in Semi-arid and sub-humid regions*. Paper presented at Symposium Environmental Sciences in Developing Countries, February 11-23. Nairobi.

Mitchell, J.K., J.C. Branch, and E.R. Swanson. 1980. "Costs and Benefits of Terraces for Erosion Control", *Journal of Soil and Water Conservation*. Vol. 35, No. 5, September-October.

Mitra, T. 1981. "Some Results on the Optimal Depletion of Exhaustible Resources Under Negative Discounting", *Review of Economic Studies*. Vol. 48, pp. 521-532.

Moore, T.R. 1978. *An Initial Assessment of Rainfall Erosivity in East Africa.* Technical Communication No. 11, Dept of Soil Science, University of Nairobi, Kenya.

Moore, T.R., D.B. Thomas, and R.G. Barber. 1979. "The Influence of Grass Cover on Runoff and Soil Erosion from Soils in the Machakos Area, Kenya," *Tropical Agriculture*. Trinidad Vol. 56, No. 4 October pp. 338-344.

Moorhead, H.J. and G.P. Sims. 1982. "Sediment Deposition in Reservoirs in the River Tana, Kenya," Paper presented to the 14th Congress of Large Dams, Rio de Janiero. Proceedings pp. 601-613.

Morgan, K.M. and R. Nalepa. 1982. "Application of Aerial Photographic and Computer Analysis of the USLF for Area Erosion Studies," *Journal of Soil and Water Conservation*. Vol. 37, No. 6 November-December.

Morgan, R.J. 1965. *Governing Soil Conservation - Thirty Years of The New Decentralization*. Baltimore, Johns Hopkins Press.

Morgan, R.P.C. ed. 1980. *Soil Conservation, Problems and Prospects*. New York, John Wiley and Sons.

Moroccan National Man and the Biosphere Committee. 1979. "Rapport National Marocain," *Hommes, Terre et Eaux*. Vol. 9, No. 30 January-February, pp. 27-54.

Mougenot, F. 1972. "L'eau et le Sel dans l'Agriculture Marocaine," *Hommes, Terre et Eaux*. No. 3-2, pp. 51-65.

Murray, G.F. 1979. *Terraces, Trees and the Halian Peasant, An Assessment of Twenty-five Years of Erosion Control in Rural Haiti*. Prepared for USAID/Haiti, October.

Murray, G.F. 1982. *Cash-Cropping Agroforestry in An Anthropological Approach to Agricultural Development in Rural Haiti*. Paper presented at Wingspread Conference on Haiti, Present State and Future Prospect, September 20, Racine, Wisconsin.

Murray-Rust, D. 1972. Soil Erosion and Sedimentation in a Grazing Zone West of Arusha, Northern Tanzania. *Geografiska Annaler*. 54A 3-4, 325-344.

Musgrave, G.W. 1954. "Estimating Land Erosion - Sheet Erosion," *Publication No. 1*. Association Internationale d'Hydrologie Scientifique, Assemble Generale de Rome.

Musro, J.C. 1977. "Soil degradation in the Republic of Argentina", (112) FAO Soils Bulletin #34.

Muthoo, M.K. 1976. "Economic Evaluation of the environmental Effects of Erosion," *Agriculture and Environment*. Vol. 3, pp. 21-29.

Nadir, M.I. 1970. "Some Approaches to the Theory and Measurement of Total Factor Productivity, A Survey," *Journal of Economic Literature*. Vol. 8 No. 4 December pp. 1137-1177.

Narayana, D.V.V. and R. Babu. 1983. "Estimation of Soil Erosion in India," *Journal of Irrigation and Drainage Engineering*. Vol. 109, No. 4, December, pp. 419-436.

Narayanan, A.S., M.T. Lee, and E.R. Swanson. 1974. "Economic Analysis of Erosion and Sedimentation, Lab Orchard Lake Watershed," *Agricultural Economics Research Report 128*. 11EQ Document 74-29 Agricultural Experiment Station, University of Illinois of Urbana-Champaign/State of Illinois Institute for Environmental Quality, August.

Narayanan, A.S., M.T. Lee, K. Gunterman, W.D. Seitz, and E.R. Swanson. 1974. "Economic Analysis of Erosion and Sedimentation, Memdota West Fork Watershed," *Agricultural Economics Research Report 126*. 11EQ Document 74-13 Agricultural Experiment Station, University of Illinois, Urban-Champaign/State of Illinois Institute for Environmental Quality, April.

National Association of Conservation Districts. 1980. *National Agricultural Lands Study, Interim Report Number Four, Soil Degradation, Effects on Agricultural Productivity*. Washington, D.C. N.P. November.

National Association of Countries Research Foundation. 1980. *Disappearing Farmlands*. Second edition, August.

Neher, P.A. 1976. "Democratic Exploitation of a Replenishable Resource," *Journal of Public Economics*. 5, pp. 361-371.

Nelson, D.O. *et al*. 1980. *A Reconnaissance Inventory of The Major Ecological Land Units and their Watershed Conditions in Nepal*. WP/17 FO,NEP/74/020 Integrated Watershed Management, Torrent Control and Land use Development Project Katmandu, Department of Soil Conservations and Watershed Management/FAO/UNDP, February.

Nelson, M. 1983. "Economics of Land Clearing," *AGREP Division Working Paper No. 79*. World Bank, October.

Nelson, M. 1984. "Economic Issues in Sedimentation," *Water International*. Vol. 9, No. 4, December.

Nelson, R. 1991. "Managing Drylands", *Finance & Development*. Vol. 28 1 Mar pp. 22-23.

Nielsen, A.D. 1963. *Economics and Soil Science - Copartners in Land Classification*. Paper presented at the Region 7 Land Classification Meeting, February 12.

Nobe, K.C. and D.W. Seckler. 1979. "An Economic and Policy Analysis of Soil- Water Problems and Conservation Programs in the Kingdom of Lesotho," *LASA Research Report No. 3*. Maseru, Lesotho, Ministry of Agriculture, Kingdom of Lesotho, Fort Collins, Colorado, Department of Economics, Colorado State University, September.

Nor, S.M. and S.P. Francis. 1983. "Conservation in Malaysia," *The Planter*. Vol. 59, No. 692, November, pp. 483-490.

Norgaard, R.B. 1983. "Coevolutionary Potential," *Working Paper No. 245*. California Agricultural Experiment Station, Gianni Foundation of Agricultural Economics, Berkeley, University of California, January.

Noronha, R. 1982. *Seeing People for the Trees, Social Issues in Forestry*. Paper Prepared for Conference on Forestry and Development in Asia. 19-23 April, Bangalore, India.

Norton, R. and L. Soils eds. 1983. *The Book of CHAC, Programming Studies for Mexican Agriculture*. Baltimore, The Johns Hopkins University Press.

Odell, R.T. 1950. "Measurement of the Productivity of Soils Under Various Environmental Conditions," *Agronomy Journal*. Vol. 42, No. 6 June pp. 282-92.

O'Riordan, T. 1977. "Environmental Ideologies," *Environment and Planning A*,. Vol. 9, pp. 3-14.

Ogg, C.R., L. Christensen, and R. Heimlich. 1979. "Economics of Water Quality in Agriculture A Literature Review," United States Department of Agriculture, E.S.C.S. -58. July.

Ogg, C.R., R. Heimlich and J. Hasteller. 1980. "A Modelling Approach to Watershed Conservation Planning," *Journal of Soil and Water Conservation*. Vol. 35, No. 6 November-December.

Okuguchi, K. 1979. "Technical Progress, Population Growth and Intergenerational Equity in a Model with Many Exhaustible and Renewable Resources," *Economic Letters*. Vol. 3 pp. 57-60.

Oldenstadt, D.L. and others. 1982. "Solutions to Environmental and Economic Problems STEEP," *Science*. Vol. 217, No. 4563 September 3, pp. 904-909.

Olivry, J.C. 1977. "Transports solides er suspension au Cameroon," in UNESCO and LAHS, *Erosion and Sediment Transport in Inland Waters*. pp. 134-141.

Olson, G.W. 1973. "Improving Uses of Soils in Latin America," *Geoderma*. Vol. 9. pp. 257-267.

Olson, G.W. 1981. *Soils and the Environment - A Guide to Soil Surveys and Their Applications*. New York, Chapman and Hall.

Omar, M.H. 1980. *The Economic Value of Agrometerological Information and Advice*. WMO Report. Geneva.

Ongweny, G.S. 1977. *Problems of Soil Erosion and Sedimentation in Selected Water Catchment Areas in Kenya with Special Reference to the Tana River*. Paper Prepared for the United Nations Water Conference, March, Mar del Plata, Argentina. U. N. Water Conference E/Con. F. 70/ABSTRATES 23 June 29. 1976.

Organization for Economic Cooperation and Development. 1981. *Forestry Project Appraisal A Framework for Public Administration*. Cooperative Action Programme Joint Activity on Forestry Projects. Paris, OECD, January.

Organization of American States. 1974. *Cuenca del Rio de la Plata*. Estudio para su Platificacion y Desarollo. Washington, D.C. OAS.

Organization of American States. 1981a. *Energy and Natural Resources Activities of Technical Cooperation*. Department of Regional Development. Washington, D.C. OAS, August.

Organization of American States. 1981b. *Results of Technical Cooperation*. Department of Regional Development, Energy and Natural Resources. Washington, D.C. OAS, August.

Oswald, E. 1978. "The Surface Water Quality Impacts of Resource Management Plans, A Structure for Analysis," United States Department of Agriculture Economic Research Service Natural Resource Economics Division. *Working Paper No. 48*. April.

Oswald, E. 1981. "Biota Quality, A Riparian Habitat Model," United States Department of Agriculture Economics and Statistics Service. *Staff Report No. AGESS 810611*. June.

Oyegun, R. 1983. "Erosion-active Surfaces on a Pediment Slope," Trinidad, *Tropical Agriculture*. Vol. 60 pp. 53-55.

Pacheco, B., J. Cancio, and H. Rea. 1979. "Repuesta de Melilotus Albus a la Inocculacion, en el Certro-Este de Santiago de Estero area El Colorado, Republica Argentina," *IDIA*. No. 379-384 July-December, pp. 97-102.

Page, T. 1977a. *Conservation and Economic Efficiency, An Approach to Materials Policy*. Baltimore, The Johns Hopkins University Press.

Page, T. 1977b. "Discounting and Intergenerational Equity," *Futures*. Vol. 9, No. 5 October, pp. 377-382.

Page, T. 1977c. "Equitable Use of the Resource Base," *Environment and Planning A*. Vol. 9, pp. 15-22.

Paraguay, Government of. 1982. *Propuesta de Solicitual de una Cooperacion Tecnica para un Proyecto de Conservacion de Suelos en el Paraguay*. Asuncion, Ministerio de Agricultural y Ganaderia, August.

Parkey, P.S. 1982. "Soil Productivity, Falling or Rising? Letter. *Agricultural Engineering*. Vol. 63, No. 9 September pp. 7.

Parrington, J., W.H. Zoller, and N.K. Aras. 1983. "Asian Dust, Seasonal Transport to the Hawaiian Islands," *Science*. Vol. 20, pp. 195-197.

Paskett, C. 1981. *Methode D'estimation des Crues en vue du Dimensionnement des Ourage de Correction Torrentielle*. UNDP/FAO MOR 78/015. n. p. January.

Patil, R. and D. Sohoni. 1969. Long Term Economic Benefits of Soil Conservation Programme. *Journal of Soil and Water Conservation in India*. 17 1. 2, 22-26.

Pearce, D. 1977. "Accounting for the Future," *Futures*. Vol. 9, No. 5 October pp. 360- 364.

Pemberton, E.L. 1983. "Review of reservoir sedimentation for Yonki dam project," Unpublished paper.

Penta, A. and F. Rossi. 1973. "Objective Criteria to Declare a Series of Data Sufficient for Technical Purposes," in UNESCO/WMO/LAHS, *Design of Water Resources Projects with Inadequate Data*. Proceedings of the Madrid Symposium, June, pp. 227-240.

Pereira, C. 1981. "Rehabilitating Eroded Hill Lands in the Peoples Republic of China," *World Crops*. September/October pp. 96-99, 110.

Pereira, C. 1982. *Soil Aid Water Management Technologies for Tropical Forests*. Draft Report Prepared for the Congress at the United States, Office of Technology Assessment. June 8.

Pereira, Sir H.C. *Land Use and Practice in Himalayan Watersheds*. Draft Paper not dated.

Pereira, Sir H.C. 1984. "Land Use and Policy Practice in Himalayan Watersheds," Paper presented at the Agricultural Sector Symposia, The World Bank, Washington, D.C. January.

Perkins, D, and J.K. Culbertson. 1968. *Hydrographic and Sedimentation Survey of Kajakani Reservoir, Afghanistan*. Geological Survey of Afghanistan, Kabul, December.

Peters, W.B. 1973. *A Report on the Expert Consultation on Land Evaluation for Rural Purposes, Wageninger, The Netherlands October* Denver, Colorado, United States Department of the Interior, Bureau of Reclamation, March 26.

Peters, W.B. 1975. *Economic Land Classification for the Prevention and Reclamation of Salt- Affected Lands*. Paper Prepared for the Export Consultation on Prognosis of Salt-Affected Soils, FAO, June 3-6, Rome Italy.

Peters, W.B. 1977. *Views on Land Selection. for Water and Land Development*. Paper Presented to the Egypt Consultation on Land Evaluation for Irrigation sponsored by FAO, February 27-March 2, Rome, Italy.

Peters, W.B. 1979 *The Role of Water and Land Resource Information in World Bank Programs for Agricultural Development*. Paper Prepared for the Workshop on Soil Resources Inventories and Development, sponsored by the Soil Resource Study Group of the Agronomy Department, Cornell University, December 11-15, Ithaca, New York.

Peters, W.B. 1983. *World Soils Policy*. Memorandum Washington, D.C. World Bank, April 6.

Peters, W.B, and R.J. Winger. 1983. *Egypt, East Delta-Lake Manzala Project Identification*. Memorandum. Washington, D.C. World Bank, February 4.

Peterson, J. 1964. "The Relation of Soil Fertility to Soil Erosion," *Journal of Soil and Water Conservation*. 19 1, 115-119.

Pierce, F.S., W.E. Larson, R.H. Dowdy, and W.A.P. Graham. 1983. "Productivity of Soils, Assessing Long-Term Changes Due to Erosion," *Journal of Soil and Water Conservation*. Vol. 38, No. 1 January-February, pp. 39-44.

Pimental, *et al.* 1982. "Land degradation,Effects on food and energy resources", *Science,*. Vol 194, pp. 149-155.

Pinstrip, A. Per. 1981. Nutritional Consequences of Agricultural Projects, Conceptual Relationships and Assessment Approaches. *World Bank Staff Working Paper, No. 456*. Washington, D.C. World Bank.

Pobedimsky, A. 1973. "Relations Between Project Economics and Hydrological Data," In UNESCO/WMO/IAHS, *Design of Water Resources Projects with Inadequate Data*. Proceedings of the Madrid Symposium, June, pp. 683-696.

Portch, S. and J.L. Hicks. 1980. "A Soil Conservation Program for Ecuador," *Journal of Soil and Water Conservation*. September-October. pp. 243-244.

Portney, P. ed. 1978. *Current Issues in U.S. Environmental Policy*. Baltimore, The Johns Hopkins University Press.

Prato, T. 1987. "Allocation of Federal Assistance to Soil Conservation," *Land Economics*. Vol. 63 2 May pp. 193-200.

Prieto B.J. 1951. "Vision General de Problema de la Erosion en Colombia," *Agric Trop. Bogota*. Vol. 7, No. 9 p. 52-54.

Prospero, J.M., R.A. Glaccum, and R.T. Nees. 1981. "Atmospheric Transport of Soil Dust from Africa to South America," Nature, 289 February 12, pp. 570-572.

Purdue University. 1976. *Soil Erosion, Prediction and Control*. The proceedings of a national conference on soil erosion, May 24-26. West Lafayette, Indiana.

Pushparajah, E. Lock, Chin Sien eds. 1980. *Proceedings of the Conference on Soil Science and Agricultural Development in Malaysia*. Kuala Lumpur, Malaysian Society of Soil Science.

Pushparajah, E. 1983. "The Need for Soil Conservation in Malaysia, *The Planter*. Vol. 59, No. 692, November, pp. 513-517.

Quintela-Gois, C. 1973. "Some Criteria Used in Hydrologic Studies with Inadequate Data," In UNESCO/WMO/IAHS, *Design of Water Resources Projects with Inadequate Data*. Proceedings of the Madrid Symposium, June, pp. 241-252.

Raeder-Roitzsch, J. and A. Masrur. 1969. "Some Hydrologic Relationships of Natural Vegetation in the Chirpine Belt of West Pakistan," *The Pakistan Journal of Forestry*. 19 1, 81-98.

Raitt, D. 1981. "A Computerized System for Estimating and Displaying Shortrun Costs of Soil Conservation Practices, United States Department of Agriculture, Economic Research Service. *Technical Bulletin Number 1659*. August.

Rajan, B.H. 1982. "Reservoir Sedimentation Studies of Tungabhadra Reservoir Project, Karnataka, India," Paper presented to the 14th Congress of Large Dams, Rio de Janiero.

Ramakrishnan, S. compiler. 1978. *Land Policies in Developing Countries, Select Bibliography on Land Reform*. 1973-1971 Lincoln Institute Monograph #78-5, Cambridge, Mass, Lincoln Institute of Land Policy, August.

Ramsay, D.M. 1978. *Implications of Adopting a Land-Use Policy Based Exclusively on Land Capabilities*. Project Working Paper Jan/78/006, Rome, FAO.

Ramsay, W. and C. Anderson. 1972. *Managing the Environment, An Economic Primer*. Basic Books, Inc. Publishers, New York.

Ramsey, F. 1928. "A Mathematical Theory of Savings," *The Economic Journal*. 38, pp. 543-559.

Randall, A. and J. Stoll. 1980. "Consumer's Surplus in Commodity Space," *American Economic Review*. Vol. 70, pp. 449-55.

Ranganathan, S. 1981. *The Cost-Benefit of Conservation*. Paper prepared for the seminar "Approaches to a Conservation Strategy for Gujarat", January 21-22, Sasan Gir, Junagadh District.

Rapp. A. 1977. "Soil erosion and reservoir sedimentation - Case studies in Tanzania", In *Soil conservation and management in developing countries.* FAO Soils Bulletin #33, pp. 123-132.

Rauschkold, R.S. 1971. "Land degradation", FAO *Soils bulletin series,* No 13 FAO, Rome.

Repetto, R. 1990. "Wasting Assets, The Need for National Resource Accounting," *Technology Review.* Vol. 93 1 Jan pp. 38-44.

Research Directorate, Department of Environment, Housing and Community Development. 1978. "Economic Evaluation of Eppalock Catchment Soil Conservation Project, Victoria," Commonwealth and State Government Collaborative Soil Conservation Study 1975-77, *Report 9.* Canberra, Australian Government Publishing Service.

Reynolds, L.G. ed. 1975. *Agriculture in Development Theory.* New Haven, Yale University Press.

Ribaudo, M.O. 1986. "Consideration of Offsite Impacts in Targeting Soil Conservation Programs," *Land Economics.* Vol. 62 4 Nov pp. 402-411.

Rijsberman, F.R. and M.G. Wolman, eds. 1984. *Quantification of the effects of erosion on soil productivity in an international context.* Delft Hydraulics Lab (Netherlands).

Ritler, J.R. 1977. *Reconnaissance of Sediment Transport and Chemical Morphology in the Lower Rio. Bermejo Basin, Argentina - with a Section on Reconnaissance of the Lower Rio Pilcomayo Basin, Argentina and Paraguay.* U.S. Geological Survey, Open-File Report 76-564 Washington, D.C. Department of Interior.

Robillard, P.M.W and L. Bruckner. 1981. *Planning Guide for Evaluating Agricltural Nonpoint Source Water Quality Controls.* Athens, Georgia, United States Environmental Protection Agency, Environmental Research Laboratory.

Robillard, P.M.W. and R. Heiem. 1980. "Evaluation of Agricultural Sediment Control Practices Relative to Water Quality Planning," *Journal of the Northeastern Agricultural Economics Council.* Vol. 9, No. 1 April.

Robinson, A.R. 1981. "Erosion and Sediment Control in China's Yellow River Basin," *Journal of Soil and Water Conservation.* Vol. 36, No. 3 May-June pp. 125-127.

Roehl, J.W. "Sediment Source Areas, Delivery Ratios and Influencing Morphological Factors," *Extract of publications No. 59.* I. A. S. H. Commission on Land Erosion, pp. 202-213.

Rohdenburg, H. 1977. "Beispiele for Holozane Flachenbildung in Noral und Westafrika," *CATENA.* Vol. 4 June 65-109.

Roose, E.J, J. Godfrey, and M. Muller. 1975. "Estimation of losses of fertiliser elements by runoff and drainage in the soil of a banana plantation in the south of the Ivory Coast," (Fre) *Fruits.* Vol 30 (4) pp223-235, April.

Roose, E.J. 1976. "Use of the Universal Soil Loss Equation on Predict Erosion In West Africa," in Soil Erosion, Prediction and Control, paper to a National

Conference on Soil Erosion, Soil Conservation Society of America, May 24-26, pp. 61-65.

Roose, E.J. 1977. "Application of the Universal soil loss equation of Wischmeier and Smith in West Africa", pp. 177-188, in Greenland and Lal.

Rostow, W.W. 1960. *The Stages of Economic Growth*. Cambridge, Cambridge University Press.

Runge, C.F. 1987. "Induced Agricultural Innovation and Environmental Quality, The Case of Groundwater Regulation," *Land Economics*. Vol. 63 (3) Aug pp. 249-258.

Runge, C.F and D. Halbach. 1990. "Export Demand, U.S. Farm Income and Land Prices, 1949-1985," *Land Economics*. Vol. 66 2 May pp. 150-162.

Russel, E.W. 1980. *Soil Conditions and Plant Growth*. London, Longman.

Rutherberg, H.V and C. Lehman. 1980. "The Economics Loss through Soil Erosion (An Example)," *Quarterly Journal of International Agriculture*. Vol. 19, No. 3 July - September, pp. 300-303.

Sabherwal, B.K. 1973. "Designing Projects for the Development of Groundwater Resources in the Alluvial Plains of Northern India on the Basis of Inadequate Data," in UNESCO/WMO/IAHS, *Design of Water Resources Projects with Inadequate Data*. Proceedings of the Madrid Symposium, June, pp. 365-381.

Sampath, R.K. 1990. "Some Aspects of Irrigation Distribution in India," *Land Economics*. Vol. 66 4 Nov pp. 448-463.

Sampson, N.R. 1982. "U.S. Lands and Water Resources, Demands in Conflict," *Agricultural Engineering*. Vol. 63, No. 2 February pp. 20-21.

Sanders, D. and W.M. Stevens. 1981. *Institutional Requirements for the Development of Land Conservation Programmes*. Paper Prepared for Informal Workshop on the Economic Aspects of Land Conservation under Humid and Semi-Humid Tropical Conditions, June 30-July 3, Rome.

Sassone, P. and W. Schaffer. 1978. *Cost Benefit Analysis, A Handbook*. New York Academic Press, Inc.

Sastry, G and V.V. Dhruva Narayana. 1984. "Watershed Responses to Conservation Measures," *Journal of Irrigation and Drainage Engineering*. Vol. 110, No. 1 March , 14-21.

Schertz, D.L. 1983. "The Basis for Soil Loss Tolerances," *Journal of Soil and Water Conservation*. Vol. 38, No. 1 January-February pp. 10-14.

Schmitz, A. and D. Seckler. 1970. "Mechanized Agriculture and Social Welfare, The Case of the Tomato Harvester," *American Journal of Agricultural Economics*. Vol. 52, pp. 569-77.

Schramm, G. 1979. *Subsoiling in Mexico-Boon or Boondoggle? Some Perverse Economics/Technology Literature*. Paper Presented at the 18th Annual Meeting of the Western Regional Science Association, February 23-25. San Diego, California.

Schramm, G. 1982a. *Monitoring the Benefits and Costs of Integrated Rural Development Projects, Jamaica*. Paper Prepared for the 21st Annual meeting of the Western Regional Science Association, February 27. Santa Barbara, California.

Schramm, G. 1982b. "The Economics of Soil Conservation in a Semi-Arid Country, Mexico," Mimeo. School of Natural Resources, University of Michigan, May.

Schumer, S.A. 1963. "The Disparity between Present Rates of Denudation and Orogery," U.S. Geological Survey, *Professional Paper No. 454H.*

Science & Education Administration, USDA. 1979. *Animal Waste Utilization on Crop- Land and Pastureland, A Manual for Evaluating Agronomic and Environmental Effects.* USDA Utilization Research Report No. 6, United States Environmental Protection Agency - 600/2-79-059.

Scitovsky, T. 1941. "A Note on Welfare Propositions in Economics," *Review of Economic Studies.* Vol. 9, pp. 77-88.

Seginer, I. 1966. "Gully Development and Sediment Yield," *Journal of Hydrology.* Vol. 4, No. 3, pp. 236-53.

Seitz, W.D., C.R. Taylor, R.G. F. Spitze, C. Osteen, and M.C. Nelson. 1979. "Economic Impacts of Soil Erosion Control," *Land Economics.* Vol. 55, No. 1 February pp. 28-42.

Seitz, W.D. and E.R. Swanson. 1980. "Economics of Soil Conservation from the Farmers Perspective," *American Journal of Agricultural Economics.* Vol. 62, No. 5 December pp. 1084-1088.

Seldon, T.H., L.D. Walker, Pa Mong, and Mun-Chi-Yong. 1968. *Economic Evaluation and Selection of Land for Irrigation.* Paper Prepared for the Soil Survey Seminar, Department of Land Development, Ministry of National Development, August 15, Bangkok, Thailand.

Senanayake, R. 1983. "The Ecological, Energetic and Agronomic System of Ancient and Modern Sri Lanka," *The Ecologist.* Vol. 13, No. 4, pp. 136-140.

Sfeir-Younis, A. 1982. *Economic Aspects of Soil Conservation Programs in LDC's.* Paper Prepared for the International Symposium and Exhibition "Polders of the World", October 4-10. Lelystad, The Netherlands.

Sfeir-Younis, A. 1983. "Economic Aspects of Soil Conservation Programs in Less- Developed Countries LDC's," *Water International.* Vol. 8, pp. 82-89.

Sfeir-Younis, A. 1984a. "Aspectos Economico en la Evaluacion de Poro Yectos de Conservacion de Suelos," Paper presented at the U. Federico Marroquin. Guatamala City, Guatamala. May 9-11.

Sfeir-Younis, A. 1984b. "The Management of Sediments in Developing Countries, A Socioeconomic Perspective," Unpublished draft. August.

Sfeir-Younis, A. 1984c. "Sedimentation Management in Developing Countries, Socioeconomic Issues," Paper presented at the East-West Center, Conference on Sedimentation of Dams. May 12-18. Honolulu, Hawaii.

Shah, S.L. "Ecological Degradation and Future of Agriculture in the Himalayas," *Indian Journal of Agricultural Economics.* pp. 1-22.

Shalash, M. and E. Salah. 1977. "Erosion and Solid Matter Transport in Inland Waters with Reference to the Nile basin," in UNESCO and IAHS, Erosion and Solid Matter Transport in Inland Waters, pp. 278-283.

Shaler, N.S. "The Economic Aspects of Soil Erosion," *National Geographic,.* Vol. 7, No. 10 October 1896 pp. 328-338.

Shaller, D.V. 1981. "The Sociology of a Stove," *Unasylum.* Vol. 33, No. 134 pp. 30-33.

Shannon, R. 1975. "Simulation, A Survey with Research Suggestions," *AIIE. Transactions.* Vol. 7, p. 289-301.

Sharma, H.D. and H.R. Sharma. 1977. "Sediment Problems at Intakes for Hydropower Plants," in UNESCO and LAHS, *Erosion and Solid Matter Transport in Inland Waters.* pp. 330-337.

Sharma, H.R. 1973. "Sediment Problems of Hydropower Plants", in International Area of Hydrologic Research, *Paper and Proceedings.* 15th Congress. Istambul, Turkey, Vol. I, pp. A67-1 to A67-7.

Sharma, R. 1983. "The Green Revolution is Depleting our Soil Micronutrients," *Ecodevelopment News.* No. 24-25 March-June pp. 15-16, 47.

Sharp, B.M.H. and D.W. Bromley. 1979. "Soil as an Economic Resource," *Economic Issues No. 38.* Madison, Department of Agricultural Economics, University of Wisconsin October.

Shaxson, T.F. 1981. *Introduction Paper on Land Conservation Measures and Related Costs.* Paper Prepared for Informal Workshop on the Economic Aspects of Land Conservation under Humid and Semi-Humid Tropical Conditions, June 30-July 3, FAO, Rome.

Shen, H.W. and H. Kikkawas, eds. *Application of Stochastic Processes in Sediment Transport.* Littleton, Colorado, Water Resources Publications n.d.

Shepard, R. 1970. *Cost and Profit Functions.* Princeton University Press.

Short, C. and E.O. Heady. 1983. "Programmed Interrelationships Between Soil Loss and Exports," *CARD Report No. 120.* Ames, Iowa, The Center for Agricultural and Rural Development, Iowa State University, September.

Shortle, J.S. and J.A. Miranowski. 1987. "Intertemporal Soil Resource Use, Is It Socially Excessive?," *Jrnl of Environmental Economics & Mgmt.* Vol. 14 2 Jun pp. 99-111.

Shrader, W.D. *Effects of Erosion and Other Physical Processes on Productivity of U.S. Croplands and Rangelands.* Report Prepared for the Congress of the United States, Office of Technology Assessment, n. d.

Shrader, W.D., H.P. Johnson, and J.F. Timmons. 1963. "Applying Erosion Control Principles," *Journal of Soil and Water Conservation.* Vol. 18, No. 5 September-October pp. 195- 199.

Simpson, J.R. and R.A. Young. 1973. "Open Access and the Economics of Grazing Land Conservation, The Papago Case," *Journal of Soil and Water Conservation.* August, pp. 165-168.

Sinden, J. and A. Worrell. 1979. *Unpriced Values, Decisions without Market Prices.* New York, John Wiley & Sons.

Singer, M.J., P. Jaritzky, and J. Blackard. 1982. "The influence of Exchangeable Sodiva Percentage of Soil Erodibility," *Soil Science Society of American Journal.* Vol. 46, No. pp. 117-121.

Singh, T. 1965. "Soil Conservation Can More Than Double Production," *YOJANA*. 9 24, 23-24.

Singh, T. 1979. *Preliminary Guidelines for Environmental Impact Studies for Watershed Management and Development in Mountain Area*. Draft, Rome, FAO, June.

Singh, V.P. 1982. *Statistical Analysis of Rainfall and Runoff*. Water Resources Publication. Michigan, Bookcrafters, Inc.

Skidmore, E.L., P.S. Fischer, and N.P. Woodruff. 1970. "Wind Erosion Equation, Computer Solution and Application," *Soil Science of American Proceedings*. Vol. 34 5 November - December, pp. 932-935.

Smith, E. and B. English. 1983. *Determining Wind Erosion on the Great Plains*. Center for Agricultural and Rural Development, Iowa State University, January.

Smith, S.C. and E.N. Castle. 1979. *Economic and Public Policy in Water Resource Development*. Ames, Iowa, Iowa State University Press.

Smith, V.K. 1978. "Scarcity and Growth Reconsidered," *American Journal of Agricltural Economics*. Vol. 60, No. 2 May, pp. 284-89.

Soemarwoto, O. 1974. *The Soil Erosion Problem in Java*. Paper Presented at the First International Congress on Ecology, the Hague, September. Bonding, Institute of Ecology, Padjadjaran University.

Soil Conservation Service. 1981. "Economic Effects of Land Treatment on Onsite Sheet and Rill Erosion Control," *Technical Note, Watershed Planning, Series No. 1704*. Fort Worth, Texas, South Technical Service Center U.S.D.A June 2.

Soil Conservation Service, New South Wales. 1981. *Annual Report 1981*. New South Wales, Australia.

Soil Conservation Society of America. 1979. *Soil Conservation Policies, An Assessment*. Ankenylowa, SCSA.

Sokolov, A.A., S.E. Rantz, and M. Roche. 1976. *Floodflow Computation*. The Unesco Press.

Solow, R.M. 1974. "Intergenerational Equity and Exhaustible Resources," *Review of Economic Studies*. Vol. pp. 29-45.

Southgate, D. and R. Macke. 1989. "The Downstream Benefits of Soil Conservation in Third World Hydroelectric Watersheds," *Land Economics*. Vol. 65 1 Feb pp. 38-48.

Southgate, D. 1990. "The Causes of Land Degradation Along 'Spontaneously' Expanding Agricultural Frontiers in the Third World," *Land Economics*. Vol. 66 1 Feb pp. 94-101.

Spears, J.S. 1980. "Can Farming and Forestry Co-exists in the Tropics," *Unasylva*. 32 128. pp. 2-16.

Squire, L. and H. Van der Tak. 1975. *Economic Analysis of Projects*. Baltimore, The Johns Hopkins University Press.

Stall, J.B. 1962. "Soil Conservation Can Reduce Reservoir Sedimentation," *Public Works*. September pp. 125-128.

Stamp, L.D. 1952. *Land for Tomorrow The Underdeveloped World*. Bloomington, Indiana University Press, New York, American Geographical Society.

Steele, I. 1977. "Central America Fights a Battle of Soil Erosion," *Christian Science Monitor*. March 29, p. 14.

Stevens, M.E. 1978. *Land Use Pattern for Marginal Land and Erosion*. N.p.

Stiles, D.N. 1983. "Camel Pastoralism and Desertification in Northern Kenya," *Desertification Control*. No. 8 June pp. 2-8.

Stocking, M.A. 1978. *Remarkable erosion in Central Rhodesia*. Proceedings of the Geographical Association of Rhodesia, Vol II pp. 42-46.

Stocking, M.A. 1984a. "Erosion and soil productivity, A review", *AGLS*. FAO Rome.

Stocking, M.A. 1984b. *Rates of erosion and sediment yield in the African environment*. LAHS Publication No 144.

Stocking, M.A and A. Pain. 1983. "Soil Life and the Minimum Soil Depth for Productive Yields, Developing a New Concept," *Discussion Paper No. 150*. School of Development Studies, University of East Anglia, Norwich, November.

Stoeckler, J.H. 1962. "Shelterbelt Influence on Great Plains Field Environment and Crops," *Production Research Report No. 62*. USDA, Forest Service, October.

Stoevener, H. and R. Kraynick. 1979. "On Augmenting Community Economic Performance by New or Continuing Irrigation Developments," *American Journal of Agricultural Economics*. Vol. 61, pp. 1115-1123.

Stoll, J.B. 1962. "Soil Conservation can Reduce Reservoir Sedimentation," *Public Works*. Vol. 39, No. 9, September, p. 125.

Stonehouse, D.P and M. Bohl. 1990. "Land Degradation Issues in Canadian Agriculture", *Canadian Public Policy Canada*. Vol. 16 4 Dec pp. 418-431.

Street-Perrott, A. *et. al.* eds. 1983. *Variations in The Global Water Budget*. Boston, D. Reidel Publishing Company.

Strotz, R.H. 1956. "Myopia and Inconsistency in Dynamic Utility Maximization," *Rev. of Economic Studies*. 23, pp. 165-180.

Struthers, R.E. 1981. *Economic Guidelines for Planning of Water Resource Projects*. Englewood, Colorado, PRC Engineering Consultants, Inc. August.

Struyk, R.J. 1988. "The Distribution of Tenant Benefits from Rent Control in Urban Jordan," *Land Economics*. Vol. 64 2 May pp. 125-134.

Suarez de Castro, F. 1951. "Influencia de la Erosion Sobre la Productividad de los Terrenos," *Agricultural Tropical Bogota*. Vol. 7 pp. 42-43.

Sunarno, Ir. and Ir. Sutadji. 1982. "Reservoir Sedimentation, Technical Environmental Effects. Paper presented to the 14th Congress of Large Dams, Rio de Janeiro. Proceedings pp. 489-508.

Sundborg, A. 1983. "Sediment problems in River Basins," *Nature and Resources*. Vol. 19, No. 2. April-June, pp. 10-21.

Suward Jo, Sofijaj Abujamin. "Crop Residue Mulch for Conserving Soil in Upland Agriculture," abstract *Paper Summaries, International Conference on Soil Erosion and Conservation.* p 59 (no date).

Svetlosanov, V. and W. Knisel, editors. 1982. *European and United States Case Studies in Application of the CREAMS Model.* IIASA Collaborative Proceedings Series. CP-82-S11.

Swaminathan, M.S. 1973. *Our Agricultural Future.* Sander Patel Memorial Lectures. 1973 at India International Centre, New Delhi October 30, 31 and November 1. New Delhi, All Indian Radio. n. d.

Swanson, E.R. and C.E. Harshbarger. 1964. "An Economic Analysis of Effects of Soil Loss on Crop Yields," *Journal of Soil and Water Conservation.* Vol. 19, No. 5 September-October. pp. 183-186.

Swanson, E.R. and D. MacCallum. 1969. "Income Effects of Rainfall Erosion Control," *Journal of Soil and Water Conservation.* 242, 56-59.

Swanson, E.R. 1977. *Economic Evaluation of Soil Erosion Productivity Losses and Off-site Damages.* Paper at seminar "The Economic Impact of Section 208 Planning on Agricultural," Great Plains Resource Economies Committee, June 8-9, Colorado.

Swindale, L.D. ed. 1978. *Soil Resource Data for Agricultural Development.* Honolulu, University of Hawaii.

Szechowycz, R.W. and M.M. Quereshi. 1973. "Sedimentation in Mangla Reservoir," *J. of the Hydraulics Division.* Proceedings of the American Society of Civil Engineers. September.

Taneja, D.D. "Sediment Problems Related to Increased Discharges at the Intake of a Run-of-River Canal System in Haryana India," in UNESCO and LAHS, *Erosion and Solid Matter Transport in-land waters, LAHS,.* pp. 344-352.

Tejwani, K.G. 1979. "Malady-Remedy Analysis for Soil and Water Conservation in India," *Indian Journal of Soil Conservation.* Vol. 7, No. 1, April, pp. 27-45.

Tejwani, K.G. 1984. "Reservoir sedimentation in India, Its causes, control, and future of action, Unpublished paper.

Temple, P. 1972. Measurement of Runoff and Soil Erosion at an Erosion Plot Scale, with particular reference to Tanzania. *Geografiska Annaler.* 54A 34, 203-220.

Thacher, B.S. 1979. "Desertification, The Greatest Single Environmental Threat," *Desertification Control.* No. 1, June pp. 7-10.

Thomas, M.F. and Whittington, eds. 1969. *Environment and Land use in Africa.* London, Metheun.

Thompson, M.E. and H.H. Stoevener. 1983. "Estimating Residential Flood Control Benefit Using Implicit Price Equation," *Water Resources Bulletin.* Vol 9, No. 6, December, pp. 889-895.

Thorbecke, E. ed. 1969. *The Role of Agriculture in Economic Development.* New York, Colombia University Press.

Trimble, S, and S.W. Lund. 1982. "Soil Conservation in the Coon Creek Basin," *Journal of Soil and Water Conservation*. Vol. 37 No. 6 November - December pp. 355-356.

Trolh, F., J. Hobbs and R. Donahue. 1980. *Soil and Water Conservation for Productivity and Environmental Protection*. Englewood Cliffs, N. J. Prentice-Hall, Inc.

Turner, S. 1982. "Soil Conservation, Administrative and Extension Approaches in Lesotho," *Agricultural Administration*. Vol. 9 pp. 147-62.

UNESCO. 1980a. "Casebook of Methods of Consumption of Quantitative Changes in the Hydrological Regime of River Basins due to Human Activities," *Studies and Reports in Hydrology No. 28*. UNESCO, Paris.

UNESCO. 1980b. *Casebook of Methods of Computation of Quantitative Changes in the Hydrology of River Basins due to Human Activities*. Paris.

UNESCO. 1982. "Sedimentation Problems in River Basins," *Studies and Reports in Hydrology*. No. 35 Edited by A. Sunbong. Paris, France.

UNESCO. 1984. *Implementation of the Second Phase of the International Hydrological Program, Report of the Secretariat*. Draft report. Paris January.

United Nations. 1976. *Guidelines for Flood Loss Prevention and Management in LDC's*. U. N. Department of Social Affairs, National/Water Series No. 5, U. N. New York.

United Nations. 1977. "Assessment of Water Resources, Networks, Surveys and Services Related Facilities, Present Status and Requirements by 2000," Prepared jointly by WMO and UNESCO for the *UN Water Conference*. Mar del Plata, Argentina March.

United Nations Conference on Desertification. 1977. *Transnational Greenbelt in North Africa Morocco - Algeria - Tunisia - Libya - Egypt*. Background Document. Nairobi, Kenya, United Nationals A/CONF. 74/25 August 29 - September 9.

United Nations Conference on Water. 1973. *Plan D'Action de Mar del Plata*. Mer de Plata, Argentina, March 14-25.

United Nations Development Programme/U.N. Food and Agricultural Organization. 1974. *Upland Development and Watershed Management Republic of Korea Comprehensive Technical Report*. AGL,DP/ROK/67/522 Technical Report 1 Rome, FOA.

United National Economic Commission for Asia and the Far East. 1953. "The Sediment Problem,. *Flood Control Series*. No. 5, ST/ECAFE/SER. F/5 Bangkok, ECAFE.

United Nations Economic Commission for Asia and the Far East. 1955. "Multiple - Purpose River Basin Development", Part 1. *Manual of River Basin Planning. Flood Control Series*. No. 7, St/ECAFE/SERF/7, New York, ECAFE.

United Nations Economic Commission for Asia and the Far East. 1964. "Manual of Standards and Criteria for Planning Water Resource Projects," *Water Resources Series No. 26*. ST/ECAFE/SER. F/26, New York, United Nations.

United Nations Economic Commission for Asia and the Far East/United Nations Office of Technical Cooperation. 1969. "Planning Water Resources Development," *Water Resources Series No. 37*. St/ECAFE/SER. F/37 n. p. ECAFE.

United Nations Environment Program. 1981a. *Consultative Group for Desertification Control*. Summaries of Project Proposals DESCON -3/2 June.

United Nations Environment Program. 1981b. "Cost-Benefit Analysis a Tool for Sound Environmental Programme," Special issue of Industry and Environment No. 2, p 1-22.

United States Army Corps of Engineers. 1970. *Proceeding of a Seminar on Sediment Transport in Rivers and Reservoirs*. Davis, California, The Hydrologic Engineering Center.

United States Army Corps of Engineers. 1982. "Engineering and Design Draft Chapters of Sedimentation Manual," Circular No. 1110-2-241. Washington D.C. Department of the Army November.

United States Department of Agriculture. 1981. *Soil, Water and Related Resources in the United States*. Analysis of Resource Trends. Washington D.C. USDA, August.

United States Department of the Interior. *Bureau of Reclamation Manual Vol. 5 Irrigated Land Use,*. Part 2 Land Classification.

Uribe Henao, A. 1971. "Erosion y conservation de Suelos er Cafe y Otros cultivos," *Cericafe,*. Vol 22 No 1. Jan-Feb pp1-17.

Van Wambeke, A. 1981. "Calculated soil moisture and temperature regions of South America", SMSS Technical Monograph No 2, SCS, USDA November.

Walker, D.J. 1982. "A Damage Function to Evaluate Erosion Control Economics," *American Journal of Agricultural Economics*. November.

Walton, A.L. 1981. "Intergenerational equity and resource use", *The American Journal of Economics and Sociology,*. Vol 40 (3) July, pp. 239-248.

Ward, R. 1978. *Floods, A geographical perspective*. MacMillan Press, New York.

Willis, K.G, Benson, J.F. and C.M. Saunders. 1988. "The Impact of Agricultural Policy on the Costs of Nature Conservation," *Land Economics*. Vol. 64 2 May pp. 147-157.

Wischmeier, W.H. 1976. "Use and Misuse of the Universal Soil Loss Equation", *Journal of Soil and Water Conservation*. 31, 5-9.

Wischmeier, W.H and D.D. Smith. 1978. *Predicting rainfall erosion losses - A guide to conservation planning*. USDA, Agriculture Handbook No537.

WMO. 1982. *Cost benefit assessment of hydrological data*. Final draft report, Geneve, November.

World Food Programme. 1984. *Interim evaluation summary report*. Various. 1980-1984.

Wu, Deyi. 1984. "Sedimentation problems in water conservancy in China," Paper presented to the Environment Policy Institute, East-West Centre workshop on the Management of River and Reservoir Sedimentation in Asian Countries, May 14-19. Honolulu, Hawaii.

Yamauchi, H and Onoe, H. 1983. "Analytical Institutional Economics of Water and Environmental Conservation," *Water International*. Vol. 8, pp. 133-139.

Young, D. and D. Walker. 1982. "The Impact of Agricultural Technical Progress in the Long- Run Benefits of Soil Conservation," Paper Presented at the SCS Conservation Economics Workshop, Lincoln, Nebraska, April.

Yugian, Long and Zharg Quishun. 1981. "Cost allocation in water resources development. *Water Supply and Management*. Vol 5, No 4/5, pp. 351-360.

About the Book and Authors

Soil erosion has become a problem of crisis proportions in developing countries around the globe, and better approaches to land management are desperately needed. This book provides analytical frameworks to guide the creation and appraisal of soil erosion control programs. The authors discuss a broad range of important issues involved in designing and implementing more effective soil conservation programs.

The book begins with a discussion of the physical and economic dimensions of soil erosion and an estimation of the extent of the problem. It then explains how to improve the evaluation of soil conservation programs by integrating scientific knowledge with economic methods and procedures. A series of practical illustrations graphically demonstrates the application of the concepts derived from the theory.

The most important organizational, institutional, technical, and macroeconomic aspects of soil and land management appraisal are outlined, and the major elements of project or program design are also pinpointed and illustrated. Finally, the book suggests areas most in need of future research.

Alfredo Sfeir-Younis is chief economist with the Economics and Policy Division of the World Bank. **Andrew K. Dragun** is senior lecturer in law and economics at La Trobe University, Australia.